Make:

Robot
Magic

BEGINNER ROBOTICS FOR
THE MAKER AND MAGICIAN
by Mario Mar

Make:
Robot Magic

By Mario Marchese

Published by Make Community, LLC
150 Todd Road, Suite 100, Santa Rosa, CA 95407

Make: books may be purchased for educational, business, or sales promotional use.
Online editions are also available for most titles.
For more information, contact our corporate/institutional
Sales department: 800-998-9938

Publisher: Dale Dougherty
Editor: Patrick Di Justo
Creative Direction/Design: Juliann Brown
Photography: Katie Rosa Marchese
Illustrations: Mario Marchese

October 2021: First Edition
Revision History for the First Edition
9/15/2021

See www.oreilly.com/catalog/errata.csp?isbn=9781680457124 for release details.

O'Reilly Online Learning

For more than 40 years, www.oreilly.com has provided technology and business training, knowledge, and insight to help companies succeed.

Our unique network of experts and innovators share their knowledge and expertise through books, articles, conferences, and our online learning platform. O'Reilly's online learning platform gives you on-demand access to live training courses, in-depth learning paths, interactive coding environments, and a vast collection of text and video from O'Reilly and 200+ other publishers. For more information, please visit www.oreilly.com

How to Contact Us:

Please address comments and questions concerning this book to the publisher:

Make: Community
150 Todd Road, Suite 100, Santa Rosa, CA 95407

You can also send comments and questions to us by email at books@make.co.

Make: Community is a growing, global association of makers who are shaping the future of education and democratizing innovation. Through *Make:* magazine, and 200+ annual Maker Faires, *Make:* books, and more, we share the know-how of makers and promote the practice of making in schools, libraries and homes.

To learn more about *Make:* visit us at make.co.

Homer Liwag

FOR A WRITER, OR ANY CREATOR...

staring at that blank page at the start of a new endeavor is an ominous and daunting thing. But what an amazing opportunity it represents! That blank page is a chance to affect an art form, to influence new ideas, to represent a unique outlook.

Creating something from nothing is like starting with a blank page! Wonderful!

Unfortunately, much of magic is simply copying, replicating what has come before. And when that happens, the art doesn't move forward. It's frustrating to watch an art form so rich... fall into a state of stagnation through repetition .

Mario's work creates opportunity. The opportunity to inspire people to break free from stagnation and to *create*. It's an opportunity to take what's come before, and expand it into new directions. That's an incredible gift.

In my own work, if you look closely, I've made an effort to create illusions and magic using materials that the audience is familiar with. This is important.

Apart from a few exceptions, it was never about glossy red boxes with dragon patterns on them. It was about finding materials and objects that people can relate to. Be it steel scaffolding, muslin stapled to wood, light towers like those you'd see at a rock concert, or a chair on top of a table in an attic.

Mario gets that. Everything he presents is firmly tethered in the human world. A cardboard box. A piece of felt. The wire from a clothes hanger. Why is that important? Why is it better? If the audience already feels comfortable with the objects being used, the magic is ten times stronger. When something unexpected and unusual happens to those simple, recognizable objects ,the impact is greater. It affects an audience's very understanding of their everyday environment. A connection is made.

Creating magic that is as emotionally resonant as music and art has been my own life's mission. *Heart must come first.* Mario's work clearly creates from that same central source.

We see his core motivation and purpose. Heart. Family. Connection.

What you are about to read will take you through a journey of ideas. It will lead you down a path of creativity and a new look at the world and objects around you. It will remind you of the joy this art form can bring.

Accept this wonderful gift.

Take these projects and run with them.

— David Copperfield

INTRODUCTION

Welcome to a world where robots and magic intertwine! My name is Mario, but my audiences know me as "Mario the Maker Magician." I am a touring performer with a repertoire full of original robot magic, mostly created with the family audience in mind. Within the pages of this book are time-tested projects that I have developed and used in both live performances and lecture presentations all over the world. But the concept behind these developments — the combination of technology and magic — is not new. Cutting edge technology has *always* been a tool used by the magician. Often, before new technology becomes mainstream or commonly known, it is used in magic... the "unknown and new" is something that magicians have taken advantage of throughout history. Think about the first time a human voice was transmitted wirelessly. Magic! Think about the great inventor Nikola Tesla controlling a remote control toy boat in New York City's Central Park for the first time in human history. These moments scream magic in every way, and you can imagine the gears that would start turning in the minds of magicians as each new technological development came about. A few years ago, after a performance of mine, a child came up and exclaimed with enthusiasm, "My teacher told me that magic is just undiscovered science!" And I yelled back, "That's exactly right!"

My journey with robotics in magic started with one simple motivation: the desire to be able to physically walk away from my magic. It's the idea of something mechanical or robotic that can perform autonomously for a live audience, where I, the human magician on stage, become a sidekick or partner to the performance. *Robot magic.* To invoke a laugh. To create a moment of wonder. To spark curiosity. This one basic concept has long inspired my builds and still constantly motivates me. For years, I looked with awe at the legacy of Jean-Eugène Robert-Houdin, one of the most important figures in magic history. He was a French magician in the 1800s, with a background in watchmaking, who created elaborate clockwork magic automata. You may have seen a nod to his "Marvelous Orange Tree" in the 2006 film, *The Illusionist*. The original was a beautiful display of sleight of hand and audience interaction starting with a handkerchief borrowed from a spectator and leading eventually to a small, bloom-less and fruitless orange tree that magically sprouted blossoms then real oranges right before the very eyes of the spectators. The finale included two mechanical butterflies emerging from the tree with the spectator's borrowed handkerchief unfolding as they rose in the air. Robert-Houdin's genius was that he not only created these beautiful and mysterious, autonomously moving objects, but he also mixed classic sleight of hand with them. He interacted with them on stage, and made them into full theatrical performances... experiences to be had.

Over the years, and with the aid of modern DIY electronics instead of clockwork, I've been able to make many robotic creations that have helped me to accomplish that goal of mine. I was even able to create a humble cardboard* robot homage to that classic Robert-Houdin blooming orange tree. I find the reward of this work to be immense. It's one thing to complete a robotic project and watch it move. *It's another thing to take the movements of a robot to create real emotion in a live audience.*

*You'll see that cardboard is a recurring theme in my work and in this book. Why? For me, my builds are always intended for family and children audiences, and I want those kids to see my work and realize that they can do the same and make their imagination come to life, too! I always prefer humble materials that can be obtained inexpensively or even found right in the recycling bin. I choose software that is free and easy to learn. I use electronic components that are inexpensive. And yes, we'll talk about 3D printing in this book, but there are more and more household 3D printers available these days that don't cost much more than a common paper printer, and more importantly, more and more schools and libraries are offering the use of 3D printers in their spaces. The point is, accessibility in making is really important to me. So, for me, cardboard will always be king!

One of my favorite creations for my theater show is a robot monkey named Marcel. He's powered by lithium batteries and moves via servomotors controlled by an Arduino microcontroller board. Marcel has just one job during our performance: to pick up a ball and put it in a cup. For three minutes I talk to and argue with this robot on stage as he evades his job... he drops the ball, knocks the cup over, tries to eat it, and generally makes a fool out of me. In the end, after much back and forth (and an apology from me for raising my voice,) Marcel finally picks up the ball and places it in the cup, met by much celebration from the audience.

It's three minutes of theater to a live audience using a robot that I can completely walk away from. Giving him identifiable features and a name helps me introduce an immediate emotional connection to the audience. The fact that he has an extremely simple problem to solve lays the groundwork for comedy and dialogue. All ages can understand and connect with the simple premise of putting a ball in a cup. We laugh because Marcel makes it so complicated. And this is why I wanted to write this book. It's about enhancing the stories we tell, making people laugh, and creating magic and comedy using simple, barebones DIY electronics.

You might have dabbled with programming in the past and lost interest. It can be overwhelming! There are so many tutorials, so many different kinds of boards, and so many different ways to accomplish the same task through code. I'm here to break it down to its bare bones for you. If you are new to programming and robotics, this book is for you! If you know a lot about robotics and are here for new ideas and creative inspiration, this book is for you, too! Our projects will be simple in terms of programming and electronics but deep in terms of audience impact and performance tips and tricks. We will take it step by step with each project. We'll direct you toward which board to use, what servos to pick up, the easiest code to hack and understand, and more. My goal is to show you how a simple servomotor can create a whole show's worth of magic.

Let's do this!

NOTE: As you work through this book, visit mariothemagician.com/robotmagic for supplementary resources, including project code, video clips, and more!

#1
THE TOOLS TO ELEVATE WHAT YOU CREATE

The "brains" of most of my signature robotic magic routines is the Arduino UNO, a credit card sized microcontroller board that you can easily program to control all kinds of electronic components. In other words, the Arduino can help you make things move! Without any understanding of complicated electronics, you can upload prewritten code to the board and control, for example, a small servomotor. Through the Arduino, you can tell the servo to turn to a certain position, and you can tell it *when* to do so. You can ask the servo to stay still for 15 seconds, then move to 90 degrees, wait another 30 seconds, and move again to its original position at 0 degrees. It's a simple movement and so easy to create, but what can we do with that in a performance setting? You'd be surprised at the many possibilities that one simple movement can unlock.

Adobe Stock-Eva Kali

Here's one example:

Say you have a whiteboard, with magnetic letters that spell out MAGIC SHOW, resting on a table next to you. You smile at your audience, then unexpectedly sneeze. As you do so, the "M" in MAGIC SHOW falls off. This simple but effectively goofy moment can be created easily with a single servomotor that has a magnet attached to its end. Your audience assumes the whiteboard is magnetic, because they recognize the colorful plastic letter magnets that adorn so many family's refrigerators. Familiarity is a powerful suggestive measure used in magic. But in reality, the board itself is not magnetic at all, and all of the plastic letters are glued onto the board, except for the "M," which is held on thanks to the magnet attached to a servo behind the board. Hidden behind the board is also an Arduino, powered by a 9V battery. As you place the sign on your table, you turn on the Arduino. You'll have programmed the Arduino to wait, say, 30 seconds before telling the servo to move, which will release the magnetic bond and cause the letter "M" to fall. So, you have 30 seconds from when you turn the board on to when you have to make your sneeze. There's wiggle room, though! Let's say you miss the mark and sneeze at 27 seconds. Three seconds later, the letter falls off. That may be even funnier! Even if the letter falls off *before* you sneeze, it can still play well. And the truth is, every great performance has an "out" for all possible situations.

It may be assumed that working with electronics in a performance setting would be a recipe for rigidity. But in actuality, there's a lot of flexibility and fluidity in how you interact with the electronic elements. You learn to *play* with the "song" of the electronic actions, and as you practice, you'll notice something amazing. With each performance, you will innately begin to *know* exactly when to land that sneeze.

What you do after the initial sneeze/letter-fall moment is where you let your character grow, breathe, and shine. You can create an entire scene, a whole routine, a creative mini-play. You can program your servo to move as many times as you'd like. Maybe you pick up the fallen letter and place it back in position, but as you walk away the letter falls again. That's funny! We program the Arduino to move the magnet away from the letter at 30 seconds, then have it move back to its original position a few seconds later, ready for you to place the "M" back on the board. This sequence can repeat again and again. The actions taking place on the whiteboard are merely opportunities for your character to react in a funny way, whether that's in a grumpy or deadpan manner, or silly and slapstick, or however you most enjoy performing. You can create multiple storylines with a single servomotor and some simple code, and as we move through the projects in this book, you'll see exactly how to do that.

Working with robotics in magic opens up unique possibilities. Robotic sleight of hand has some advantages over the limitations of the typical human hand. Robotic sleight of hand also presents plenty of its own limitations and challenges, but let's start here with an advantage. Let's compare a cardboard hand to a human hand, for example. You could carefully conceal a magnet between cardboard layers in the cardboard hand. Concealing a magnet within a human hand would take more work (or surgery!) Take that magnetic cardboard hand and add a few servo movements, and with some creativity, you could create really powerful magic.

Your robotics don't necessarily need to be built or conceived entirely from scratch, either. I encourage you to look at everything available to you as you develop your creations. You could, for example, purchase an inexpensive robotic arm kit, then attach your magnet-infused cardboard hand to the end of the robotic arm. Take a commercially available magic trick like, say, a simple penny-to-dime trick that involves the removal of a metal "shell" to change a penny into a dime. (If you're a magician, you probably have tons of stuff like this laying around, ready to be revisited through a robotic lens. If you're new to magic, a quick request for a penny-to-dime trick at your local magic retailer or online shop will have you covered in no time.) It wouldn't be difficult to program your robotic arm with a magnetic cardboard hand to perform the penny-to-dime trick, with the aid of its magnetism. And then rather than focusing on the sleight of hand yourself, you can focus on creating interesting interaction between you and the robotic arm as it performs. Create the illusion of conversation between you and the robot. Command it to move. Interact with it. Commit the timing to memory, so when the arm rises, you can steal the shell (the sneaky part we don't want our audience to see) as the arm reaches to give you a high five! This is just one conceptualization, but the point is that perhaps a once-forgotten trick can become completely modern and brand new again through DIY electronics.

It's just like the music we adore and specific melodies and chords that we love. If we can maintain those melodies and chords, but slightly alter the way they sound, the music becomes new again. Think of how different a song can feel when played on an acoustic guitar versus a distorted electric guitar. The same goes for fashion or any kind of artistic expression. Often, the most innovative designs are those that take something that is familiar and recognizable in some way and reinvents it through a new and unexpected lens. The goal for us with robotics and magic is to build on the shoulders of classics, then try to adapt some type of robotic component to enhance, change, or replicate it. Approaching the art form in this way is much less daunting than thinking of it as a task that requires building something completely new from scratch. And that may come as well, eventually. But it's not ever where we begin. As you experiment in the merging of these two worlds, the world of robotics and the world of magic, both worlds will become new again.

One of my hopes is that after reading and tackling some of these projects, you will have new eyes. I hope that from here on out you might open magic books and think, "Could I make a robot do this?" Or, "Is there a battery-powered toy that I could adapt to replicate this trick?" And on the other hand, I hope you look at robotics or electronic devices and toys differently from here on out. I hope you see the limitations of a robotic device and think about how you could innovate within those very limitations. I hope that you start to see particular movements and imagine what magic trick might work within them. And it's not about proving that robotic-infused magic is *better* than non-robotic magic. This is an exploration of a conversation that two seemingly very different worlds might be able to have together.

THE ARDUINO

Over the years, I have created a short list of go-to electronic components that are especially useful in the kind of things I create. These components have helped me become a better magician and performer as well as a better maker. We'll use some of these components in the projects within this book. There are a few extras included in my list, though, for you to keep in mind as you delve further on your own.

As I've already mentioned, the **Arduino UNO** is probably the most important item on that list. The Arduino is the simplest and most affordable way to control or build almost anything electronic. I can't emphasize enough how important it is to get started with it. Once you upload some code and blink an LED for the first time, once you see the words and numbers you type turn into flashing lights and moving motors, you will see everything electronic differently. You will realize that anything that is battery powered can be hacked somehow with an Arduino. Even a car! Maybe an Arduino-powered automobile that finds a selected playing card? Ha! But let's not get carried away just yet. First, you'll need to learn how to set up and plug in your board, how to upload your first sketch, and how to plug in and interact with various components. In the next chapter, you'll learn all of that.

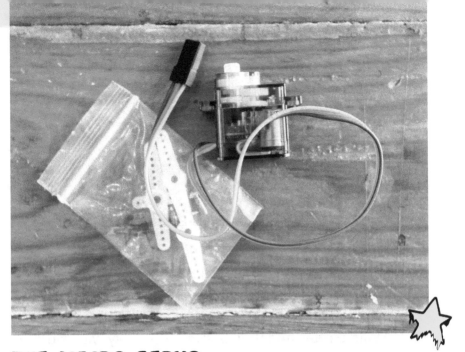

THE MICRO SERVO

Moving on in my list of go-to components, we come upon the **micro servo**. The micro servo is the type of motor I use most often with the Arduino. As we've briefly touched on already, this small, incredibly useful tool and its simple movements are perfect for creating great comedy and magic. I have built signature routines using only micro servos! They are small, affordable, very easy to program with the Arduino, and they don't use a lot of power. You can power an Arduino and a servomotor with a single 9V battery! So what does a servo do exactly? It has a little plastic arm called a servo horn and it can be programmed to move that tiny arm from 0-180 degrees. This means you could attach it to the lid of a small cigar box to open it on command! You could attach a light object, like a sponge clown nose, to a piece of wire to make it pop up or appear from behind your suitcase. You can program up to twelve servos with one Arduino UNO. That's a lot of clown noses! We'll go into detail later on about how to control this kind of movement with an Arduino, but for now, let's continue down the list of my favorite components.

THE RELAY

Next up, let's talk about **relays**. I love relays! 5V relays are easy to solder and easy to use. If you have an electronic object with an on/off switch, you can replace the physical switch with a relay and then activate the relay with the Arduino. This is an excellent tool for electronic toy hacking! You could, for example, find a simple common electronic toy like one of those little barking dogs, that toy dog that takes few steps, then stops and barks, then repeats. This could be a great little toy for magic, especially if we could control when, exactly, the dog moves. A relay will allow you to do that. This can open a lot of possibilities! Imagine having someone select a playing card. You shuffle the pack and place a line of cards on the floor, face down. The dog moves on command and stops at the selected card. A relay is also is a great way for you to control bigger motors too. Because the Arduino runs on 9V, it is unable to power larger motors alone. But the relay helps us solve that problem by connecting those larger motors to a separate battery. With a larger motor, for example, we could create a custom version of a classic card fountain and control it with a relay and an Arduino. (For the non-magic-initiated, a card fountain is a device that rapidly spits playing cards high in the air, useful for several different effects in magic and a fun, unexpected spectacle to catch your audience off-guard.)

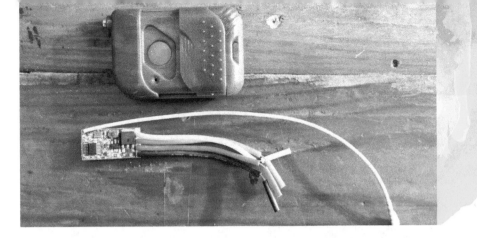

THE RC TRANSMITTER AND RECEIVER

Let's move on to the **RC (remote control) transmitter and receiver**. The kind I'm referring to is a small and inexpensive key fob with a button and an accompanying circuit that is powered by a 9V battery. The circuit gets connected to the device you want to control remotely, and the fob syncs to allow you to control the circuit remotely using the fob. With magic and performance in mind, one major benefit to these fobs is that they are small, compact (easy to conceal!) and they provide a whole other method to communicate and create magic. Some RC transmitters and receivers have relays built inside too. This means you can wirelessly control almost any toy or gizmo that was not originally intended to be remote control, opening up a lot for the creative thinker and magician. It's an exciting feeling to browse the toy section at Target with the knowledge that you can hack any of those electronic toys. RC circuits are accessible and affordable. And remote control is magic! Remember, in 1898, Nikola Tesla was the first to create wireless remote control, and he demonstrated it by powering a toy boat. It was pure magic then, and although the concept is no longer new, and it can absolutely still be used to create pure magic now.

THE DC MOTOR

Finally, we must include **DC motors** on our fundamentals list. DC motors are the type of small motors used in most cheap toy cars and trucks, because they create a *spinning* motion. Any small battery-powered toy with wheels or that spins most likely has a DC motor inside. Small DC motors often power up with under 9V (and sometimes as little as 1.5V) of battery power. Like everything else on our list, there is so much you can do with them! Remember the card fountain we mentioned back in the section about relays? If you've ever seen a magician create an impressive stream of playing cards flying up out of a top hat or box, they were most likely using a card fountain.

With the projects in this book, we will be purposefully programming electronics to create laughter. I feel this is important to keep in mind, because when you have an ultimate goal or motivation, a purpose to drive you, the work of learning how to use all these various electronic components becomes much easier. To make someone laugh or smile is an amazing thing. Even if it's just a chuckle, laughter can instantly break awkwardness and create a moment of connection between people. In fact, laughter itself should be on our list of components. It is, after all, one of the most powerful tools of all.

Once you've gained a decent grasp of the technical aspects of the electronics, then you can go on and dive into classic principles of comedy with those new eyes. Go back to Charlie Chaplin and his amazing antics. Watch Chaplin's iconic dance of dinner rolls from his 1925 film, *The Gold Rush*. Imagine using servo-powered forks to robotically replicate that beautiful scene. It's about taking the things that connect us, the scenes, tricks, gags, jokes that have proven themselves classics, and re-imagining them to make them new again. It's not about copying or stealing material. It's about taking ideas and innovating with them. It's the concept of adding servomotors and programming to reinvent something that will instantly spark an emotional connection for people. The emotional connection comes from a sense of familiarity, even if it's a vague familiarity. The spark of curiosity and joy comes from experiencing that familiarity in a brand new way.

That's my job here… to show you how to connect things that were not necessarily intended to connect, but when connected, can create something awesome. Innovation. "To mix ideas that have never been mixed before," I say that at the end of every show I perform. We must experiment and think outside of ourselves as much as we can. We must learn how to use as many tools as we can along the way. The more tools we know how to use, the more chances we have to create what we imagine, and *imagination is everything*.

"Imagination is more important than knowledge. For knowledge is limited to all we know and understand, while imagination embraces the entire world, and all there ever will be to know and understand." — *Albert Einstein*

"Don't let anyone rob you of your imagination, your creativity, or your curiosity. It's your place in the world; it's your life. Go on and do all you can with it, and make it the life you want to live." — *Dr. Mae Jemison*

Imagination *is* everything. I hope you use these small tools to create what you imagine. I hope you use them to bridge the gap between your ideas and tangible reality. And after you've built your basic foundation in robot magic by working through this book, there are so many places to look for further inspiration. On the performance end, there are countless books out there about magic tricks and performance. Some magicians say, "If you don't want people to know the secret of your trick, put it in a book!" Ha! That means those books are truly full of gems hidden in plain sight. Go through as many old magic books as you can find. Trick after trick, ask yourself what parts of your new DIY robot magic knowledge could be adapted?

And for those of you who are already well-read magicians, I encourage you to seek out more books in the maker realm and deepen your understanding of robotic and electronic possibilities. Find books like *Making Things Talk* by Tom Igoe. Search makezine.com for tutorials and more advanced projects. Dive into other boards like the Raspberry Pi. Personally, I always try to add one small challenging element to each new project I undertake. I challenge you to challenge yourselves, too! Don't stop here. There are so many different kinds of relays! So many different kinds of motors that do so many different kinds of things. And don't get me started with sensors! That's a whole other book!

If the goal is to pursue a life filled with happiness, then chase what makes you happy. If what you create makes you *and* others happy, then I'd say you've found your purpose in life. Create songs that make us all dance, make magic that inspires us to be more child-like. Enhance the story you tell to help crush the fears in others. Don't ever underestimate a project that can make someone laugh. Don't ever underestimate something that makes people believe in something bigger, something stronger within themselves. That's the magician's purpose, isn't it? They don't fool the audience for the sake of fooling the audience. They fool them so that the audience can feel like children again, so they can laugh and forget the worries of the day. The world sometimes disregards the performer as "non-essential," but in reality, they offer something incredibly vital in this world. So with each small project you tackle in this book, I hope you do make people laugh, but even more importantly, I hope you will believe in yourself a little more.

#2
GETTING STARTED WITH ARDUINO

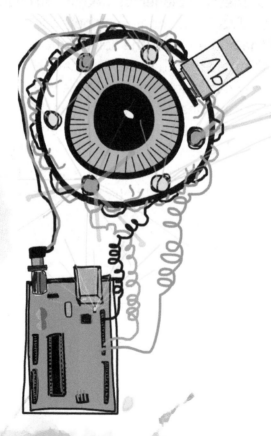

For every project that involves specialized movement, we will need something that can store the instructions for those movements and activate them. We can easily connect a small motor to a battery and watch it spin. We unplug the battery, and the motor stops. But what if we need the motor to turn on after 20 seconds without us touching it? What if we want to control the exact length of time that motor will spin? Or maybe our project needs an LED light that blinks a certain number of times for us. All of these problems require something that can remember a sequence or pattern. This is where the Arduino UNO board comes in.

The Arduino UNO can do a lot, but to keep things simple we will look at it like this: the UNO is a credit card-sized circuit board that plugs into our computer via USB. You type in code that programs the UNO, dictating how long you want things to turn on and off... things like servomotors, LEDs, and relays. You upload that code and save it to the board. The board remembers the on/off patterns you create and will play your sequence the moment you connect it to a source of power. You can add wait times in between the on/off portions of your code. You can instruct the board to play your sequence in a loop over and over or just one time. Once the code has been uploaded to the board, you can disconnect the UNO from your computer and place the board inside your project!

But let's say that once the Arduino UNO is installed in your project, you find the timing is off a little with one of the servomotor movements in your routine. The

Adobe Stock-Eva Kali

Arduino makes tweaking your code so easy. We just plug the Arduino back into your computer, adjust the code a little, unplug, and try again. This can be done thousands and thousands of times with a single board. Upload and adjust until you get things perfect.

Do you see where we are going here? See why I still get excited about this even after all these years? The automata of the past required complicated clockwork to create delays and to activate movement. The Arduino is like a brain, a flash drive storing your routine and playing it on command. You connect it to a 9V battery and it all just works! Now, we might need a second battery or a few extra components to help us along the way, but the basics are all here. Step by step, we will walk through it all together.

DOWNLOADING THE ARDUINO SOFTWARE & GETTING IT RUNNING

Before we get into the specifics of what is on the board itself, let's get the software up and running. Go to www. arduino.cc/en/software and find your computer's operating system in the download options. Arduino is compatible with both Mac and PC. After the download is complete, open the Arduino application. The window that appears with some text is called your "sketch." All of your code will be written inside the sketch. When the window first opens, you'll see:

```
void setup() {
}


void loop() {
}
```

sketch_jul16a

```
void setup() {
  // put your setup code here, to run once:

}

void loop() {
  // put your main code here, to run repeatedly:

}
```

For your sketch to work, you need both **void setup()** and **void loop()**. The moment the Arduino is turned on, everything written within the brackets of the void setup() section runs once, and then everything within the void loop() brackets runs. The void loop section will run in a loop, repeating over and over, like a song playing on repeat. In other words, we set up first, and then we sing! But the best way to clearly understand it is to see an example sketch.

LED BLINK CODE

Go back to your sketch window in the Arduino program on your computer, *select and delete everything in your current sketch* — don't worry, we'll be replacing those vital void brackets — then type the following (also found at mariothemagician.com/robotmagicchapter2):

```
/*
    Blink
    Turn on an LED for one second, then off for one
    second, repeatedly.

    This example code is in the public domain.
*/

int led = 13;        // "led" can be changed to
                     // any name



// the setup routine only runs once
void setup() {
                    // Here we set up our "led"
                    // digital pin as an output.
    pinMode(led, OUTPUT);
}

// the loop below runs over and over again

void loop() {

digitalWrite(led, HIGH); // turn the LED on
                         //(HIGH is the voltage
                         // level)

delay(1000);             // wait for a second
digitalWrite(led, LOW);  // turn the LED off by
                         // making the voltage LOW
delay(1000);             // wait for a second
}
```

Great! Now, in the program, you'll see a circle button with
a check mark in the top left corner. Click that, and save
your sketch as **LED blink 1**. If everything was typed out
correctly, your sketch will now compile. If you receive an

error instead, go back and make sure everything is copied exactly from the text and try again.

UNDERSTANDING THE CODE

Now that your sketch is compiled, let's break down the code for you to understand.

Any light grey text you see in your sketch is a "comment," which is there for your reference only and will be ignored by the Arduino board. Comments are a chance to make notes in between the code. I use comments to help me remember what a certain line of code means without affecting the code itself. Single line comments can be created by typing **//** at the beginning of a line. Everything after that will turn to light grey. You can use **/*** and ***/** to create a multi-line comment.

Next in the code, you'll see **int led = 13;** — this refers to pin 13 on the board, and in this sketch, pin 13 has been named **led**. The Arduino UNO has a built-in LED right on the board that we can use! This is great for testing code without plugging in additional components.

Next comes **void setup()**. Within the **void setup() { }** brackets, we place code that will run once, only at startup. That means it runs once when we either plus in the Arduino or when we press the reset button on the board. The setup is like an introduction before the full routine.

Within the brackets of **void setup()** in this sketch, we have one line of code that says **pinMode(led, OUTPUT);** — here we are simply telling the Arduino that the pin named led is going to output electrical current. In other words, with power, the LED will turn on. We have to set these things up first in setup, so when the **void loop()** code begins, the Arduino knows which pins to use.

```
// the loop below runs over and over again

void loop() {

digitalWrite(led, HIGH); // turn the LED on
                         //(HIGH is the voltage
                         // level)
delay(1000);             // wait for a second
digitalWrite(led, LOW);  // turn the LED off by
                         // making the voltage LOW
delay(1000);             // wait for a second
}
```

Next, we see **void loop()**. All code within the curly brackets will be repeated continuously forever, as long as the Arduino has power. The first line is **digitalWrite(led, HIGH);** — this tells the board to turn the pin named led on. As we saw in setup, the pin named led is connected to number 13 on the board, so the board gives power to pin 13 when the command **digitalWrite(led, HIGH);** is written. The next command is **delay(1000);** — delay means to wait, and 1000 means 1000 milliseconds, the equivalent of one second. You can change the number inside the delay to the amount you need. If you want the LED to stay on for five seconds, for example, write **delay(5000);** in the code. 13 seconds would be **delay(13000);** — but what if we want it to stay on for just half a second? Well, that would be **delay(500);**

Now, if the only commands inside void loop were **digitalWrite(led, HIGH);** and **delay(500);** the LED would turn on and never turn off. That's why we have **digitalWrite(led, LOW);** in the next line. This tells the board to turn the LED off.

The next command — **delay(1000);** — creates another wait time of one second. And, after the last command, the code jumps back to the top of the loop and repeats. So, if you look at the code, you can see the LED will: turn on, wait for one full second, turn off, wait for another full second, and repeat.

UPLOADING THE CODE TO YOUR BOARD & BLINKING YOUR LED

Let's get this uploaded and prove that it works! On your Arduino UNO board, there is a USB port. This takes a USB 2.0 Type A/B, the same type of USB cord generally used for household paper printers. This will connect your board to your computer.

Next, we have to tell the Arduino software that there is a board plugged in. In the software, click Tools > Board and select **Arduino UNO**. Then, click Tools > Port and select the port that has Arduino listed. It's usually the last one on the list. After both board and port are properly selected, we can upload our code by clicking the circle button with the arrow pointing right. That's the upload button. If everything has lined up correctly, your sketch will say "done uploading." Look at your board! Is the light blinking on and off steadily? You did it!

Now, play with what you have created! Change the code! Try shortening or extending the delay time numbers, and upload again, so you can start to get a feel for the control you have. The more you play, the more you'll get comfortable and familiar with the technology. The goal is for the technology to be second nature, so the later work of adapting robotic movements and actions to comedic timing is easy. As you start to develop your own routines, you'll need to be comfortable with going back and forth between uploading to the board, testing, and then going back and tweaking your code.

MORE ABOUT THE BOARD

Let's look at the board again. I'm not going to break down every single element of the Arduino board here, but there are a few things I do want to point out.

The Arduino UNO has **13 digital output pins**. These pins are used to activate various components like your relays, LEDs, and servos. That's 13 opportunities all on one board. As you progress with this technology, you may find yourself using almost all of these pins at once on a single project!

Imagine the possibilities!

The pins labeled **A0-A5** are **analog pins**. Those are used mostly as inputs to read sensors. You can declare an analog pin as a digital pin, too.

You have a **barrel jack** to power your board with an external power supply or 9V battery.

You'll see pins labeled **GND**, which stands for ground. Everything that is plugged into an Arduino data pin (like an LED), must have its other end plugged into GND to complete the circuit.

The **Vin** pin is useful as an alternative way to power the Arduino board, particularly helpful if you don't have a 9V plug for the barrel jack. It stands for Voltage In.

The Arduino UNO also has a **reset** button. That will reset your Arduino, making it re-run the sketch from **startup()** without unplugging it all.

The long black brick in the middle of the circuit board is your **microcontroller**, the "brains" of the board. The Arduino UNO uses an **ATmega328** microcontroller.

There are also several **LEDs** mounted on the UNO. One is your power LED. Another is the LED attached to digital pin 13 that we can use to test sketches. There are also LEDs (TX and RX) that show when data is being transmitted or received.

NEXT UP

So, we've blinked an LED! We've started to see that working with the Arduino is really not so scary or difficult. Dipping our toes in and demystifying the technology is the first step, and there is quite a bit we can do with LEDs. But *movement* is the crux of our robot magic projects, so that's what we're going to do next... let's *make a motor move*, folks!

#3
MAKE A MOTOR MOVE

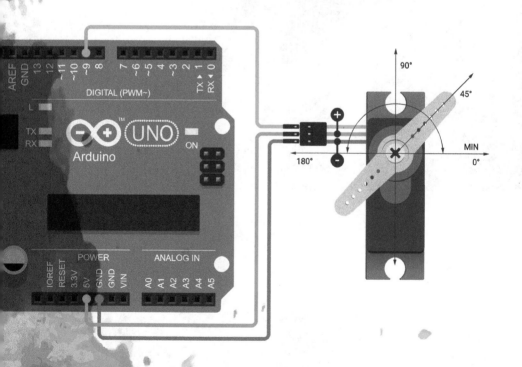

To create robotic movement, we need *motors*! But before we get into this, let's first remember that at the core, everything we're doing with the Arduino is just about switching things on and off, off and on. Whether it's a small LED that blinks, or a relay that allows us to power something bigger, it all comes back to simply turning things on and off. Working with the Arduino is just about commanding a bunch of digital pins to go on and off! Keep this in mind, and it's much less overwhelming!

In this chapter, we will discuss the easiest way to make a motor move with the Arduino. My favorite motor to use with an Arduino is a **micro servomotor**. They are cheap, accessible, and very easy to code. The best part is that they work well within the 5V that the Arduino's digital pins create. This is huge! This means you can separate the board from your computer and power both the Arduino and the micro servomotor with just a 9V battery! Think about how portable that is.

I built one project from a small coffee canister and lid. Every time I looked away, the lid would pop open and two eyeballs would appear. Then, just as fast, the eyeballs would disappear back into the can and the lid would close. Every time I looked away, it would happen again. It was so simple, but it elicited great reactions when I performed with it. That project was created with just a single micro servo and an Arduino, powered with a 9V battery. This is just one small example, but believe it or not, the foundation of almost *all* of my robot magic comes down to programming a micro servomotor with an Arduino. We will start with one motor and work up to programming multiple motors and understanding how to power them all. Through this process, we will learn the fundamentals to *animate what we make*!

WHAT IS A MICRO SERVO?

So what can a micro servo do? Don't be fooled by its size ... this little thumb-sized motor can accomplish a lot. A micro servo has a small arm that can move from 0-180 degrees. With the Arduino, you can program that arm to move to a particular position, have it stay in that position, and move again whenever you want. The servo arm has small holes in it. This is so you can attach things to it.

There are thousands of applications for servomotors and infinity ways to adapt them for performing magic and comedy. One Halloween, I built a servo-powered silly string

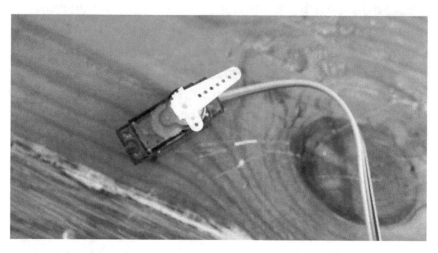

shooter (www.instructables.com/Arduino-controlled-Silly-String-shooter/) and hid it inside of a carved pumpkin. That's right, we made a puking pumpkin silly string shooter! All with a servo and an Arduino. Trick-or-treaters had a blast!

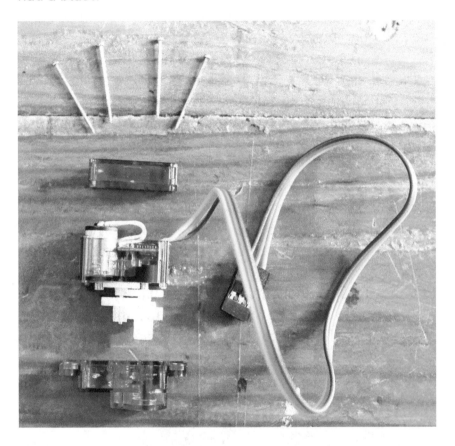

Inside a servo, we have four main components — a **gearbox**, a **potentiometer**, a **DC motor**, and a **control circuit**. The DC motor inside is very small and can power up at a high speed but is fairly low on torque, meaning it spins really fast but can't lift much. This is where the gearbox comes in. Just as the name describes, the gearbox inside is made up of little gears. The spinning gears take the motor's speed and turn it into higher torque. Think of a bicycle going up a hill. At a low gear, you are able to pedal faster, putting less pressure on the pedals to get to the top of the hill easier. It's just like that with this. The main shaft comes out of the gearbox and sticks out of the servo's plastic housing. That main shaft is where we attach the servo arm. On the underside of the shaft lies the potentiometer. A potentiometer is a variable resistor. It acts as a feedback sensor and also helps the servo stay in the position we tell it to. Try gently moving a servo arm that is programmed to stay in a certain position. It will fight you! It will always push or pull back to the position it's programmed to stay in.

But how does the servo arm know where to go? And how does it know how long to stay there? That's where the control circuit and Arduino come in. The control circuit inside of the servo reads a **PWM (pulse width modulation)** signal from the Arduino, and with this communication, all of the various components of the servo work together with the Arduino to turn on a motor, control the speed, stop it to an exact position, and hold it at that position until it's told to do otherwise. It's pretty amazing, really!

CONNECTING THE SERVO TO THE ARDUINO

You'll see three wires coming out of your servomotor. Let's connect them to the Arduino using **male-to-male jumper wires**. Plug the red wire to positive, 3.3V or 5V on the board. Plug the black wire to GND on the board, and plug and the orange/yellow wire to digital pin 8 on the board.

Now that the servo is connected to the board, let's connect the board to your computer again via USB. Then, open up the Arduino application and open a new sketch by pressing the square button that looks like a piece of paper, located right next to the circular arrow button.

Replace the code in the window with the following (also found at mariothemagician.com/robotmagicchapter3):

```
#include <Servo.h>

void setup(){

Servo servo;              // Change the name from Servo
                          // to whatever you want!

servo.attach(8);          // Here we tell the Arduino
                          // which pin "servo" is
                          // attached to
}

void loop() {

delay(1000);              // Delay just means WAIT! Remember
                          // every 1000 = 1 second

servo.write(180);         // move servo to 180 degrees

delay(1000);              // Wait 1 second

servo.write(0);           // Move servo to 0 degrees

delay(1000);
```

Now, just as when we did the LED blink sketch, go to Tools > Board and select "Arduino UNO." Then, click Tools > Port and select the correlating port (usually the last one listed.) Press the compile button (the circular checkmark.) When it's done compiling, press upload (the arrow button.) As always, if you receive an error code, double-check your sketch! The code must be copied exactly.

UNDERSTANDING THE CODE

Let's break it down, starting with **#include <Servo.h>** — we must include the **Servo.h** library here for the board to communicate with the servomotor.

Next, you'll see **Servo servo; //Change the name from «servo» to whatever you want!** As the note within the code suggests — remember, light gray text in the code is a comment for your reference only and will not be read by the Arduino — this is where we name our servo! Call it whatever you want! Say, you'd like to name your servo Meatball. That's fine! Just make sure you replace "servo" with "Meatball" *in all parts of the code*.

Next up, we see **servo.attach(8); // Here we tell the Arduino which pin # «servo» is attached to** — we need to make sure the orange/yellow servo wire is connected to the same pin referenced in the code. Here, it's pin #8. If you have the wire plugged into a different digital pin, either change the number in the code to match, or move the wire to pin #8.

Next, we see **delay(1000);** — and just like in the LED blink sketch, this tells our routine to wait! 1000 equals one second.

In the line, **servo.write(180); // Move servo to 180 degrees!,** the command **servo.write(0);** tells the servo arm to move to 0 degrees. (Change the number to change the position the arm lands on. If it said **servo.write(78);** that would move the servo arm to 78 degrees, for example.)

And then **delay(1000);** tells it to wait one second.

Do you know that between those two commands, you hold all you need to animate what you create? That's it! Once you understand these two commands, you can create all kinds of amazing magic with servomotors. You can bounce between multiple servos, too! This is a simple formula that will weed out all of the confusion that comes with coding. Again, take some time to play, attach things to the servo arm and see for yourself what the motions look like and how changing the numbers in the code affects the movements.

I use **servo.write();** for all of my coding. There are other ways to make a servo move with code! A simple Google search will show you that. But this is the way I love doing it. In my stage show, each of my robotic elements perform several-minute-long routines in front of live audiences. Through a lot of trial and error in creating all of those various routines, the thing that became most apparent is that creating a clear storyline with code is the most important thing. Overly complicated code can easily overwhelm us and prevent us from making edits and tweaks and going the extra mile to hone our routine to perfection. Having an easy-to-follow storyline in your code, through which you can program movements piece by piece, is key. I have found that using **servo.write();** is the simplest way, and that will become more clear as we work through the projects in this book. The goal of our code is to replicate human emotion with robotic movements. Those movements can demand a lot of code, sometimes 16-17 lines for just one subtle gesture. Keeping it all confined to a simple **servo.write();** and **delay();** will help you create those complicated movements with ease.

#4
GETTING STARTED WITH 3D DESIGN AND 3D PRINTING

Every trade and every art form has its tools. We use all kinds of tools every day to help create what we need in a faster and more efficient way. Think of a chef, a barber, a stonemason, a dentist, an actor, a surgeon, a landscaper, or a programmer. Think of the variety of tools each requires to best do their jobs. There are such specific skill sets and such specific accompanying tools for each path. People who make a successful career out of something have *mastered* the tools they use.

I was fortunate to have the opportunity to appear on *Sesame Street* a few years ago, and what impacted me the most during that shoot were the master puppeteers. I got to watch people like Peter Linz (Ernie on *Sesame Street*;

Walter in the 2011 film *The Muppets*) and Joey Mazzarino (Murray Monster on *Sesame Street*) as they worked, and it changed my whole perspective on puppetry. When someone like Peter or Joey holds a puppet in his hands, it becomes way more than just a tool of the trade. It becomes art — the kind of art that makes you believe in magic again. And you sit back and just *know* that what you are seeing at that moment is something that you might not ever see again. When we put all of our energy into something, the results can be life-altering. My perspective changed that day, and I am so grateful.

I know, I know, we're supposed to be introducing you to 3D design and 3D printing now! But before we jumped right into it, I wanted to help further this picture for you, this picture that tools are important not because of what they are in and of themselves, but because of what you can do with them. In *The Maker Magician's Handbook*, my intro book to magic for kids through making and crafting, I mentioned three tools, three "superpowers," that I go back to constantly and that have often helped me create my magic more efficiently and helped me bring my imagination into reality. Those three superpowers are **3D designing**, **3d printing what you design**, and my favorite, **animating what you create**. With these three, the sky's the limit! Or rather, your own imagination is the limit.

3D DESIGNING

Many times when you hack something or create something from scratch, you will find yourself in need of something custom-made for your project. Maybe a piece to fit perfectly, a holder for something specific, a way to precisely attach something to something else... that's where **3D design** comes in. It can be an amazing way to quickly prototype and test an idea you have in your head. Once you've come up with a 3D design, you upload the file to a 3D printer, which can print out an actual — you guessed it — three-dimensional version of your design!

One amazing use of 3D design is replication. Let's say you have a small R2D2 toy. You can take a digital caliper, measure out the dimensions, and replicate it on screen using a 3D design program. Let's say you want to now take that design and change it... we could make a custom R2D2 magic trick, for example! With some clever hollowing out of the base in the design, we can create an area to fit a coin and create an R2D2 "coins across" routine, a classic in magic. We 3D print our design and test it out. If it doesn't work, we can adjust measurements, tweak our design, and print again. Once you have it right, paint your print and seal it up. Now you have a completely custom one-of-a-kind prop that you can write an entire themed routine around. Just like that, an old principle of magic becomes new again.

A few of the 3D printed props from Mario's theater show

There are so many different 3D design programs out there, and everyone has different opinions on each. Some are more advanced than others, some are more affordable. The one I use is free, has lots of easy-to-understand tutorials, and it saves your work online. It's called **Tinkercad**. Go to www.tinkercad.com to sign up. Once you have registered your free account, you will see a "lessons" section to get you started. Each lesson will only take a few minutes of your time, and after following just three or four of them, you'll see how simple and easy the program is to use.

As you get into your 3D designing adventures, you don't have to have a 3D printer in your home... there are many schools and public libraries that make 3D printers accessible to their students and communities. There are even services like www. 3dhubs.com that will deliver your prints to you in days. But if you do decide to invest in a 3D printer for your home, there are many great affordable options out there to choose from! I do suggest that you invest right away in a good **digital caliper** to assist you in 3D designing.

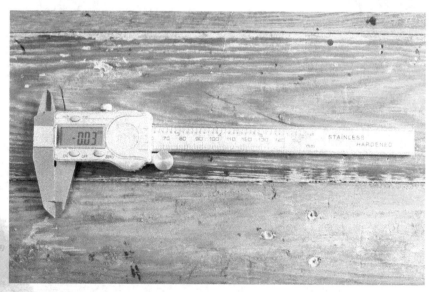

A digital caliper is a great tool for precisely measuring objects you'd like to hack or replicate digitally or for creating something to fit something perfectly. Replicating simple objects is also a fantastic way to become proficient in 3D designing. Find something handheld in your home that has a simple shape, like a die or a coin. Measure the dimensions and slowly replicate the shape on Tinkercad.

Also, check out www.thingiverse.com, the universe of things! It's an online hub for sharing 3D designs. If you have an idea for an object you'd like to make, you can use Thingiverse to see if someone else has already done the design work. You can then download the design files for free, or with an optional donation for the maker who created the thing. I've saved a lot of design time this way, finding designs on Thingiverse for super useful things like a small mount for a micro servo. And even if their design is not perfect, or not the right size, you can still download the file, then upload it to Tinkercad to adjust or resize as needed. Remember that 3D printing and designing is all about prototyping. The whole reason it exists is so we can easily make and re-make. Much of the time, I go through 15 to 20 prints that don't work before I get it just right! So don't be discouraged if the first few prints of your custom design fall flat. It's all part of the process.

My goal with my own projects is actually to use my 3D design skills and 3D printer as little as possible. They are not a first step, but rather the last option when other build methods or materials are not sufficient. My first go-to materials are always cardboard, hot glue, clothes hanger wire, paperclip wire, string, straws, cloth, magnets, and the like. Probably my favorite magic projects are the ones that look innocent and cardboard on the outside but have custom 3D printed parts hidden inside. Parts that latch other parts together. Parts that mount other parts with precision. And there's the added bonus that when your props look innocent, the expectations of your audience will be low, and then you can really wow them with unexpectedly strong magic. Also, using 3D printing only when necessary is a great way to keep your creativity flowing and your mind thinking outside the box.

3D PRINTING

There is an art to 3D printing successfully. The way a 3D printer works is like an upside-down hot glue gun with a super-thin nozzle that extrudes melted plastic. It builds your print from the bottom up, layer by layer. So if you print an object that has a piece hanging and not touching the build plate, for example, it won't print correctly. It will string up and warp your final product. As you starting printing things, you will learn to shape your objects with 3D printing in mind. Instead of sharp hollowed edges, you'll purposefully add curves in the gaps of your design, so that the upside-down glue gun nozzle will glide into each gap and be able to print it smoothly. This is something that people get really good at... designing with the limitations of printing in mind. Well-practiced 3D designers can even work within those very limitations and still create objects and toys that can spin or move right after being printed, without any post-print assembly or adjustments. You can just pop them right off the print bed and play — amazing! You can design hinges that can be printed with the pin right in place! Limitations really just mean that you need to adjust your common knowledge or thinking to make something work. Explore the designs on Thingiverse, and you will see the possibilities that come when you practice and take the time to acclimate your brain to a new way of looking at objects.

If you are like me, you can easily get obsessed with trying to be as original as possible with everything you make. In reality, we can only begin the journey of our true potential by first practicing by replicating and recreating what already exists. Even the future musician who one day will write the song that changes the world's perspective on music must start by learning someone else's song first. Don't let your pride come in the way of what your future self can one day create. Study the successful principles that work within the vision you are trying to create. Study the history behind it. There is a science and formula for every successful artistic expression, and the ones who inspire us the most are the ones who learned early on to create on the shoulders of those before us.

What inspires you? What performance or project have you seen that changed the way you interact with the world? You know... the one that forced you to smile the moment you thought of it. That's the one you build. Learn what tools are needed to recreate the project in your own way, so you can learn the process and understand it. Give credit to the original creators every step of the way. Your only goal of recreating should be to gain a deeper understanding of why it works. I spent five months learning how to sew my own jeans. I tried doing it without a pattern, and it was a terrible failure that wouldn't even fit a human body. I was so obsessed with the idea of making original jeans that I forgot that the whole concept has a traditional foundation. It wasn't until I carefully took apart a pair of jeans that fit me well already and replicated them piece by piece that I started making progress. Once I completed my first pair of jeans that actually fit, I made cardboard patterns to follow, so I would remember the process. After a few more tries, I was finally able to learn the foundation of making a pair of pants and use that foundation to finally make something that resembled my original, custom vision. And the best part was that along the way, I got new ideas that I probably never would have thought of if I hadn't journeyed through the mistakes! By the way... all the buttons? 3D printed... designed on Tinkercad!

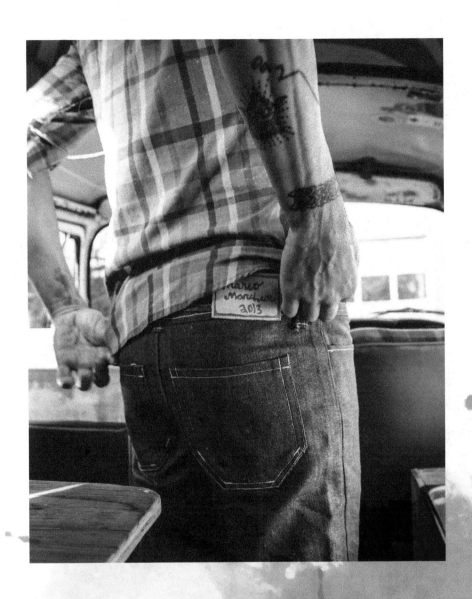

#5
SPRING SNAKE WITH BRAINS

There are some classic comedy props that we all recognize: chattering teeth, the whoopee cushion, oversized glasses, the rubber chicken, and one of my favorites, the snake-in-a-can. I love the iconic quality of each of those items. In the off-chance that you're unfamiliar with the snake-in-a-can, it's an opaque, twist-top canister, often labeled "mixed nuts" or "peanuts," that conceals a spring snake inside. It's most often used as a straightforward practical joke, in which an unsuspecting person is offered the can of nuts. Of course, when they untwist the lid, the spring snake jumps out at them and hopefully elicits a moment of surprise and laughter. The snake-in-a-can — or snake nut can, as he called it — was invented by Samuel Sorenson Adams in 1915 in response to his wife complaining about their jam jar lid not closing well. As a prank, he filled the jar with a coiled wire covered in sewn cloth, and the rest, as they say, is history. Clearly, he took a mundane life moment and ran with it. Nowadays, the snake-in-a-can gag has become so ubiquitous that few people are fooled by it. But that doesn't mean we can't still innovate with this classic. In fact, these are exactly the kind of props and principles that are screaming to be reinvented. The comedian and magician known as The Amazing Jonathan does a wonderful bit with it, completely turning the gag on its head! There's also a snake can routine in *Tarbell Course in Magic, Volume 4.* Both routines are worth looking at. Consider it an extra credit assignment!

In this chapter, we are going to gather all this inspiration to create our own version of a snake can routine I like to call the **Spring Snake with Brains**! Remember, every great old idea has the possibility to become brand new again with an Arduino and a servomotor! '

With each project in this book, I will first describe the effect or the overview of the routine. Then, we will go through the build, step by step. And finally, I will teach you how to perform it, with suggested patter and all. So, here we go...

THE EFFECT:

The performer introduces a small paper lunch bag and takes a pack of playing cards out of it. A card is selected by a spectator and then mixed back into the deck. As the cards are being mixed, the performer takes a pet spring snake out of the bag. The performer has the snake "smell" the pack before placing both snake and the cards back into the paper bag. At the performer's command, the spring snake launches out of the bag... with the selected card in its mouth!

The spring snake gag is, obviously, the most important prop for this routine. There are a lot of spring snake manufacturers, and the best ones are those where the snakes really have a good amount of spring to them. I'm using the "Fancy Salted Mixed Nuts" by the novelty and balloon company, Loftus International. The measurements in the build instructions will be suitable for that snake can. You may find the need to adjust your measurements to fit your can. Now, without further ado, let's build ourselves a servo-powered snake can gimmick!

MATERIALS & TOOLS:

- 2 identical packs of playing cards
- 2 cheap wire clothes hangers
  (like the ones you get at the dry cleaners!)
- Cheesecloth or a small piece of similar cloth
- 2 spring snake can gags
- Tupperware container lid
- Brown paper lunch bag
- Red felt
- Black felt
- Arduino UNO
- Micro servo
- Male to male wires or 22 gauge solid core wire
- 9V battery
- 9V battery clip with DC plug
- Hot glue gun + hot glue sticks
- Scissors
- Wire cutters
- Needle nose pliers
- Drill + drill bits
- Small metal file
- Masking tape
- Small Phillips head screwdriver

THE BUILD:

- We need to create a new lid for our snake can. With your scissors, cut a flat circular piece of plastic from your Tupperware lid, about the same diameter and circumference as the can (Figure ❶).

- Next, we need to make and install a hinge for our new lid. Cut a 2" square piece of cheesecloth (Figures ❷ and ❸). Hot glue a little less than one half of the cloth to the inside of the can (Figure ❹), and the other half to the outer part of your new lid, leaving roughly ⅛" of open material in between the lid and the can (Figures ❺ and ❻). This little gap will help it to open and close easily. Using thin material like cheesecloth gives the hinge a ton of flexibility without sacrificing strength. Having a lid that opens so easily helps get the most spring out of the spring snake.

- Select a drill bit that is about the same size as the gauge of the clothes hanger wire. Drill a hole in the snake can about ¼" inch up from the bottom of the can, drilling through both sides, straight across (Figures ❼ and ❽).

- Cut and stretch out the wire of the clothes hanger (Figure ❾). Thread the wire through the drilled hole, bending it upwards so it's snug against the edges of the can (Figures ❿ and ⓫).

- At one end of the wire, we need to make a bend and a loop, just above the top of the can. Make the inner loophole slightly larger than the thickness of your wire.

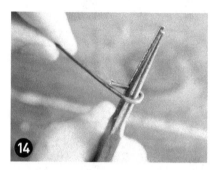

Needle nose pliers will help a lot with this. Cut the excess off with wire cutters (Figure **12**).

- At the other end of the wire, make a bend ¼" above the lid and leave a ½" tail after the bend, removing the excess with wire cutters (Figure **13**).

- Cut another piece of clothes hanger wire about 4½" long. Bend one end into a small loop (Figure **14**), but before completely closing the loop, thread it through the wire loop on the can. Then close the loop, making sure to leave enough room in the loop for your wire to hang freely. Remove any excess wire with your wire cutters (Figures **15** and **16**).

- Now, let's test the functionality of our gimmick before we finalize everything. Put your spring snake in the can and close the newly installed lid. Slide the hanging wire over the lid, securing it under the bent wire on the other side of the can. Our homemade flap door latch should securely hold the snake in place (Figure **17**).

- And now, we can use minimum movement to open the snake can. To test the action, slide the latch wire out from under the bent wire, and the snake should spring out! If the wire doesn't slide easily across the lid, play around with it, making minor adjustments with your pliers. If the lid gets stuck, use your scissors to trim it slightly until it can open freely.

- Once satisfied, press the wire tight against the sides of the can, and secure it with a few layers of masking tape at the bottom and top of the can (Figure **18**). We use masking tape so we can later adjust the wire, if need be. It's a lot easier to remove masking tape than epoxy glue!

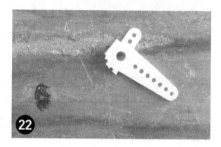

- Before we continue, use a small metal file to smooth out any sharp edges from the cut wires. We will be handling the snake can a lot after this point, so the smoother those edges are, the safer it will be to handle (Figure **19**).

So now, how do we take this mechanism and make it electronically controlled? With a few adjustments, we can have a micro servo and Arduino do all the work.

- Take a micro servo and a double-armed servo horn like the one in the photos (Figures **20** and **21**). Take your wire cutters and clip one of the long sides of the servo horn. We also need to clip one of the short sides, but it needs to be the correct side, so please be sure to refer to the photo and match it exactly (Figure **22**).

- Attach the altered servo horn to the servo. Notice the servo moves only 180 degrees. There is a stop built into the gearbox preventing the arm from traveling continuously. We have to find where the 0 or 180 mark is.

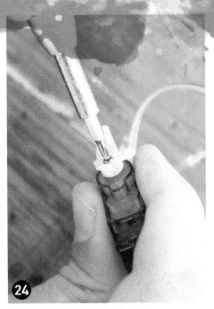

Very gently turn the arm counterclockwise until you feel it stop. Again, use the photo as your guide to match it all exactly (Figure **23**). Once it stops we know that is our mark. Remove the arm, and push it back in place in the position we just found. Use the smallest screw from the servo accessory bag to secure the servo arm in place (Figures **24** and **25**).

- With a dollop of hot glue, attach your micro servo against the can, as in the photo. Notice that we're mounting the servomotor

slightly "crooked." That angle will help secure the movement we need, so the can lid will open with minimal effort (Figures **26** and **27**).

- Create a 90 degree downward bend at the end of the latch wire, so that it just clears the in-position servo horn. After the bend, snip the tail down to ½" (Figure **28**).

- Again, let's test the movements manually to make sure everything functions well and moves easily. The goal is for that servo to move and slide the wire latch across the lid to release the snake from the can. The action should be smooth and easy. Gently test the motion by manually moving the servo arm. Make any adjustments where you need to. You might need to bend a wire here or there, or trim the lid a little.

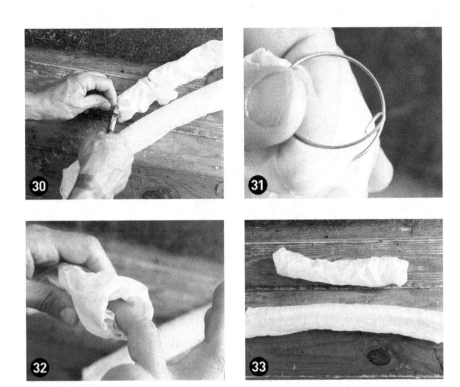

Now, let's hack the spring snake itself! We need two spring snakes for this, so take both from the snake kits in the materials list.

- With your scissors, snip one end of the fabric from one of the spring snakes (Figure ㉙). Pull back the fabric enough to find the midway point in the spring, and cut the spring in half with wire cutters (Figure ㉚). Once cut, the spring wire can be quite sharp, so take care to bend the sharp end inward with pliers (Figure ㉛).

- Tuck the excess fabric into the spring (Figure ㉜).

- We should now have one long spring snake and one half spring snake (Figure ㉝). Cut out matching red felt tongues, roughly 4" long each, and four ½" black felt circles for eyes. Hot glue them in place on both snakes,

matching them as best you can. Matching the size and placement of these adornments on the two snakes will help create a greater illusion for our magic routine (Figure **34**).

- Load the larger snake into your can with the tongue resting outside of the lid, facing the front (Figure **35**).

- For this routine, we also need two of the same exact playing card, one that our spectator will pick, and one that will later appear on the snake. With the snake loaded in the can, use just a small dab of hot glue to attach one of those playing cards to the tongue so it stays propped up and face down, as in the photo (Figure **36**).

- Now it's time to add the Arduino into the mix! Open the Arduino application on your computer, and plug your board into your computer.

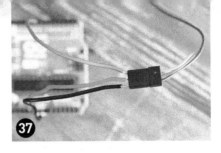

- Use male to. male wires to plug your servo to the Arduino board. With a red male to male wire, connect the red servo wire to 5V on the board. With a black wire, connect the brown servo wire to GND on the board. With a green wire, connect the yellow/orange servo wire to digital pin 8 on the board (Figures **37**, **38**, and **39**).

- In a new Arduino sketch window, replace any existing code with the following (also found at mariothemagician.com/robotmagicchapter5):

```
#include <Servo.h>

Servo snake;            //Let's call the servo snake!

void setup(){

snake.attach(8);

snake.write(160);       // move to start position
                        // before we begin

}
```

```
void loop(){

delay(10000);              // wait 10 seconds before
                           // the snake pops

snake.write(30);           // the snake pops!

delay(5000);               // wait 5 seconds

snake.write(160);          // move back to start

delay(50000);              // wait 50 seconds

}
```

• Compile and upload your code (as described in chapter two of this book.) Once it's uploaded, unplug your board from the computer.

• Load the snake into the can and gently twist your servo back into position. Slide the wire latch in place to secure the lid.

• Hold the bottom of the can against a table, and make sure the lid is not pointed at anyone! Plug the Arduino back into the computer, to give it power, so we can test our code with the snake can. Once the Arduino has power, it will start running the code immediately, so be ready! The snake will pop out after ten seconds.

• If the snake doesn't pop out, troubleshoot and make adjustments. Maybe the wire latch that secures the lid is too long? If so, trim it a bit, so that the snake releases more easily. You can also adjust the code! Change **snake.write(30);** to **snake.write(10);** That will make the arm move out farther. Change **snake.write(160);** to **snake.write(180);**. That will designate the servo's starting position closer to the can.

40

- Once you are satisfied with everything, unplug your board from the computer, attach the battery clip to the 9V battery, but leave the power plug *unplugged* from the board for now.

- Tape down your wires, and mount your Arduino and battery onto the back of the can. A loop of duct tape for the board and another for the battery works well (Figure **40**).

- Carefully load your snake back into the can, latch it, and place the whole thing inside a paper lunch bag. Push it to one side within the bag, and place your shorter snake in the bag next to the can (Figure **41**).

41

- Take the 9V power plug from your battery and rest it on the tip of the barrel jack on the board. When it's time to turn it on, you'll just have to press it in.

Let's get to the magic!

- Remember, we have two of the same playing card. One is glued to the tongue of the snake inside the bag, ready to pop out, and the matching card will be the one our spectator picks. To ensure that happens, we'll need to "force" that card. That's magician language for making someone select a specific card while maintaining their belief that it's their own free choice. And the first step in the force process we will use for this routine is to make sure the matching card is the top card in our pack. So place the card face down on top of the deck, then put the cards back in the box, and place the pack in the lunch bag.

Decorate the bag, if you'd like! Make it your own!

THE PERFORMANCE:

- Take out your bag and place it on the table in front of your audience.

- Take out the cards. Fan them out, showing the faces, and say, "I have a deck of cards, all different! You! Please take this pack and name any number between five and 25!"

- Hand them the pack, and after they name a number, have them count to that number from the top-down, placing each card one by one in a pile on the table, face down (Figure **42**).

- When they reach their number, tell them to pick up the counted pile and place it on top of the remaining pack in a criss-cross, creating a plus sign (Figure **43**).

- Now pause to reiterate what happened. You showed them a whole pack of different cards, they named *any number* they wanted. Now, have them pick up their counted pile and memorize the bottom card (Figure **44**). Guess what? You just forced a card! That's the beauty of magic. That little pause of explaining what just happened is a form of misdirection to make sure the audience forgets that the bottom card on their pile is actually the first card at the top of the pack! It will always be the

top card of the original pack, no matter what number they chose to count to. But it doesn't end there!

- They've memorized the bottom card on the pile. Now, have them place it anywhere in the deck. Have them shuffle the deck thoroughly (Figure **45**).

- While they are shuffling, tell them you have a pet! A pet snake that lives in the bag. Take the half-snake and lift it out just far enough so everyone can see its eyes and tongue. Don't lift it completely out of the bag, because we don't want to reveal how short it is (Figure **46**).

- Place the shuffled deck of cards inside of the bag next to the half-snake. Make sure no part of the snake is visible above the bag. As you place the cards in the bag, use the opportunity to push the plug into the Arduino to turn it on, without your audience knowing.

- Wiggle your fingers over the bag with both hands. Make a silly dance, shake your bottom, shout, "SPRING SNAKE POOF!" Tell your audience to shout "POOF!" When they do, snap back and say "POOF, not POOP!" All this until ten seconds pass, and POP! The snake flies out of the bag, with the selected card attached to its mouth! Show the snake, show the card, and take your bow (Figure **47**)! (And don't forget to unplug the Arduino!)

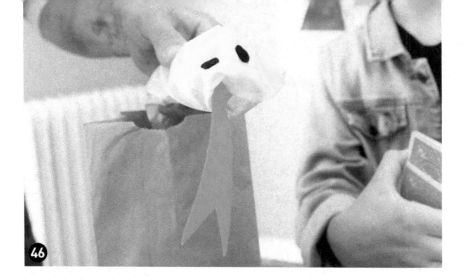

46

47

#6
BALLOON CAR CARD TRICK

For this project, we are going to take a step back from electronics for a moment to build a non-electronic balloon-powered toy car! Consider this is a little breather from the hard work of learning new technology, and a chance to play for a bit. We'll also be taking the project a step further, of course, showing you how to use that balloon-powered toy car to perform a card trick! We've created an innocent-looking craft project, but we are mixing it with magic to create an unexpected impact. Most people would not immediately associate a homemade balloon toy car with a magic prop. We're mixing worlds and creating something unique and imaginative!

MATERIALS & TOOLS: ⭐

- 2 Bic-style pens
- 2 plastic drinking straws
- 4 water bottle caps
- 1 balloon
- 1 rubberband
- 1 paper lunch bag
- **Thin cardboard that is kraft brown color on at least one side** (pizza boxes are a great source for this!)
- **2 identical packs of playing cards**
- **Hot glue gun**
- **Hot glue sticks**
- **Craft glue stick**
- **Scissors**
- **Pencil**
- **Permanent marker**
- **Ruler**

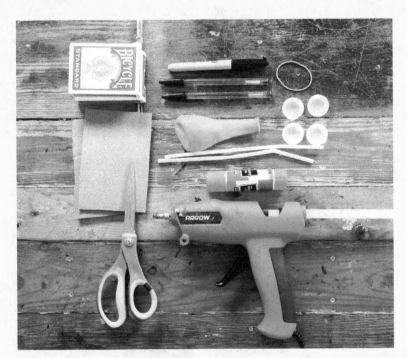

THE EFFECT:

The story goes like this. You have someone pick a card, and it gets mixed back in the deck. The cards are then spread out in a big mess on the floor or table. You introduce your toy balloon car, show it top and bottom to your audience. You blow the balloon up, put the car on the road of cards, and it zips right over the cards. Your spectator picks the car up and finds their selected card attached right to the belly of the car! Let's get rolling! Pun intended.

THE BUILD:

• Cut two matching 3½" × 5" rectangles of cardboard. At least one side of each piece should be standard Kraft brown cardboard color with no markings (Figure ❶).

• Cut two 3" pieces of straw, and hot glue one piece of straw along each short end of one of the cardboard rectangles, as in the photo (Figure ❷ and ❸).

• Remove the ink chambers from two pens (Figure ❹).

• With your scissors, *carefully* puncture a hole in the center of each of the four water bottle caps (Figure ❺ and ❻). The holes should be just large enough for the ink chamber to push through.

• Push one bottle cap onto one end of one of the ink chambers, with the inside of the bottle cap facing outward. Repeat with the other ink chamber and another cap (Figure ❼).

• Use a small amount of hot glue to secure each ink chamber on the inside of the bottle cap, keeping the cap centered until the glue dries (Figure ❽).

- Now, thread one of the ink chambers through one of the straws on your cardboard rectangle and attach another bottle cap to the other end of the ink chamber, once again securing the wheel in place with hot glue. Don't forget to hold the cap centered until the glue dries. The more centered each cap is, the better our car will work! Repeat with the other ink chamber (Figure ❾ and ❿).

- Our car is coming together, but we need something to make it move! Take the barrel from one of your pens and carefully cut a 2" piece (Figure ⓫).

- Attach the mouth of your balloon around one end of the pen barrel piece, and secure it in place with the rubber band (Figure ⓬).

- Turn your car over, so the wheels are on the ground. Secure the non-balloon end of the pen barrel to one of the short ends of the car body with a generous amount of hot glue, letting the pen barrel stick out about ¼" from the car (Figure ⓭).

- Let's test the car! Blow up the balloon and release the car on a smooth surface. Watch it zip across the table or floor!

- Now we need to create a gimmick (secret part) for the magic trick portion of our project. Turn your car upside down, wheels up. Take three different playing cards, and lay them face down and overlapping each other in a row in between the wheel axles on the car. The cards should be spread into a position to fit perfectly in between the axles *but not tucked under* the axles (Figure **14**). Glue the cards together in this position with craft glue, but do not glue the cards to the car (Figure **15**).

- Carefully peel off a thin layer of cardboard from the solid brown side of your other cardboard piece (Figure **16**).

- Glue the thin layer of cardboard to the face side of your card gimmick (Figure **17**). Trim the edges precisely (Figure **18**). You should now have a gimmick that appears to be a piece of cardboard on one side and three face-down cards on the other side.

- With your permanent marker, draw an "X" on the center of the brown side of your gimmick. Draw a matching "X" on the underside of your car (Figure **19**).

- Select two matching playing cards, one from each of your packs of cards. With a tiny dab of hot glue, secure one of the cards face up on the underside of your car, covering the "X." Keep the matching card for the next step (Figure **20**).

In final preparation to perform this routine, place the matching playing card at the bottom of a deck of cards, and place the cards back in the box. This will be our "force card," the card that we make sure our spectator chooses during the performance. Load your car and box of prepped playing cards in a paper lunch bag. Now, you're ready for an audience!

THE PERFORMANCE:

- Take your toy car out of the bag, and place it upside down, toward the side of your performance table or surface, with the cardboard shell hiding the card, and the balloon "exhaust" facing the audience. We keep it to the side for now, because we don't want to draw extra attention to the toy until we need it.

- Take your prepped playing cards out of the bag. Remember, your force card should be at the bottom of the pack. Spread a few cards from the top of the pack, and show them to your audience. Grab two or three cards at a time, going back and grabbing a few more each time, to show that all of the cards are different. Do this four or five times. The bottom card always stays face down in your other hand. As you are doing this say, "I have a bunch of different cards!"

- Square the cards, place them on the table. We're going to use a force method called the "Cross-the-cut Force" to make sure our spectator selects the right card. (This method was first published by Max Holden in Edward Bagshawe's *The Magical Monthly*, Vol. 2 No. 10, July 1925, p. 199-200.) Point to a spectator, and say, "Can you please cut these cards?" Instruct them to pick up a portion of the cards, cutting the cards anywhere in the deck that they'd like, and to place the selected portion face down right next to the rest of the cards.

- Pick up the bottom portion of cards, and place them on top of the selected portion of cards, in a criss-cross, creating a plus sign (Figure ㉑).

- Now, introduce your homemade car, lifting the car up, with the balloon tip facing up, and the balloon side of the car facing the audience. Be sure to hold the gimmicked shell firmly to the car as you lift it, so the shell doesn't move or slide out of place.

- Explain that your car is made from water bottle caps, pen pieces, pizza boxes, and hot glue. Show both sides of the car, without letting go of the shell. When showing the underside (Figure ㉒), explain that "X marks the spot!" Say it more than once! We want to be sure to draw clear attention to that X, so the audience

remembers later that they clearly saw the bottom of the car! Place the car back down on the table, upside down, with the "exhaust" facing the audience.

- Return to the spectator that cut the cards, remind them that they cut the cards wherever they wanted. Ask them to please pick up the criss-crossed portion of cards from atop the pack, look at the card on the bottom of that packet of cards, and *remember their card*.

- Then, ask them to mix up the cards thoroughly by spreading them all face down in front of them on the table, creating a long, messy strip of cards. You can help them place the cards into the correct formation (Figure ㉓).

- Next, ask your spectator to stand at the other side of the messy road of cards, ready to catch the toy car. "The toy car is going to find your selected card! Remember, X marks the spot!"

- Pinch both sides of the car as you blow the balloon up with the balloon facing the audience (Figure **24**).

- Hold the air in the balloon, and place the car on top of the other end of the messy road of cards (Figure **25**).

- Once the car is on top of the first few cards on your end of the "road," aimed toward your spectator waiting on the other end, release the shell, letting it drop to the surface below. The shell is now imperceivable to your audience, blending in with the rest of the cards (Figure **26**).

- Release the air and let the car go (Figure **27**)! The car will ride across the cards to be caught by your spectator at the other end of the table.

- "Remember, X marks the spot!" Have your spectator pick up the toy car to examine it. BOOM! Their selected card is magically pinned to the underside of the toy (Figure **28**)! You may choose to use repositionable glue to adhere the card to the car, instead of hot glue, if you'd like the spectator to be able to remove the card from the car.

- As they examine the car and react, scoop up all the playing cards in the road, along with the shell, careful not to turn it around, and shove them all in the paper bag, thus removing any evidence.

Congratulations! You just did amazing magic with a balloon-powered pizza box toy!

#7
CUBER GOOBER

The Rubik's Cube was invented by sculptor Erno Rubik in 1974, and aptly enough for our purposes, he originally called it the Magic Cube. Names aside, imagine creating a piece of art that would one day evolve not only into an iconic toy but also a worldwide sport! It's been said that it took Rubik himself an entire month to solve his puzzle upon first creating it! There are so many ways the cube can be mixed that the chance you might accidentally solve it is nearly impossible. Now, decades later, over 350 million cubes have been sold worldwide, and the world record time to solve a Rubik's cube is under four seconds! That is pretty amazing... memorizing movements and recognizing patterns of color to achieve a solution so fast.

In the magic world, the cube has become a fixture in many magicians' acts, too. Some magicians have made whole careers centered around Rubik's Cube magic! It's a powerful puzzle and a powerful prop, and with this next project, we will create a *robot* that performs with it. We will adapt traditional principles of magic with robotics as a fun exercise to see how a human and robot can work together in rhythm to fool an audience. My favorite part of this build is how compact and strong the magic is! The whole thing is self-contained, and the effect is clear and easy for an audience to follow.

THE EFFECT:

The performer shows all sides of a small, mixed Rubik's Cube and introduces a small cardboard robot. The cube is placed inside the robot. The robot opens and closes its "mouth" several times, finally revealing the cube once more, but this time it's solved! The solved cube is then taken out of the robot and shown on all sides.

MATERIALS & TOOLS:

- 4 mini Rubik's Cubes
- **1 piece of cardboard** (a pizza box is a good source!)
- **1 sheet of stiff black or dark felt**
- **1 playing card box**
- **1 index card**
- **1 roll of floral wire**
- **1 plastic drinking straw**
- **1 push pin thumbtack**
- **1 hobby X-Acto knife**
- **1 Arduino UNO**
- **2 micro servos**
- **1 9V battery plug**
- **1 9V battery**
- **6 male to male wires** (or 22 gauge solid core wire)
- **Hot glue gun + glue**
- **Scissors**
- **Painter's tape**
- **Masking tape**
- **Small screwdriver**
- **Wire cutter and stripper**
- **Pliers**
- **Ruler**

1

THE BUILD:

- Take *one* of the four small Rubik's Cubes, and mix it up (Figure **1**). (Notice that as you mix it, the center square of each side never changes. This is important to keep in mind for later.)

- After the cube is thoroughly mixed, hold it between your thumb and fingers, and turn it so you only see three sides of the mixed-up cube (Figure **2**).

- Cover those three sides with painter's tape, removing any excess tape with scissors (Figures **3** and **4**).

- With the aid of an X-Acto knife, carefully remove any stickers from the un-taped sides that *do not match* the center sticker for that side. Leave the center stickers and any stickers that match that side's center sticker in place. Be sure not to damage the stickers. Set the stickers aside for now, saving them for a later step (Figures **5** and **6**).

- Now, put the stickers back on the un-taped sides, filling in each side with just one color each, so those three sides of the cube appear to be "solved." Be careful to position each sticker as straight as possible, so nothing looks altered! (You will need to take some additional squares from a second cube to complete each of the three sides, and some of the originally peeled stickers will be left over (Figure **7**). You can discard those.)

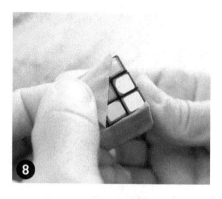

- Carefully and slowly remove the painter's tape from the other sides (Figures **8** and **9**).

Let's pause the build for a moment. Hold the cube with your thumb and fingers, so that only three mixed sides are visible to you (Figure **10**). Now, turn the cube so that only solved sides are visible to you (Figure **11**). This movement is the core of the illusion we are creating! By spinning the cube 180 degrees on its diagonal axis, you can change an apparently mixed cube to an apparently solved one. But how can we make a servo do this for us? And how can we make it look like magic? Let's get back to the build!

- Take a thumbtack and carefully twist and push the pin into the corner of the cube. Make sure the corner you select is one that simultaneously touches at least one

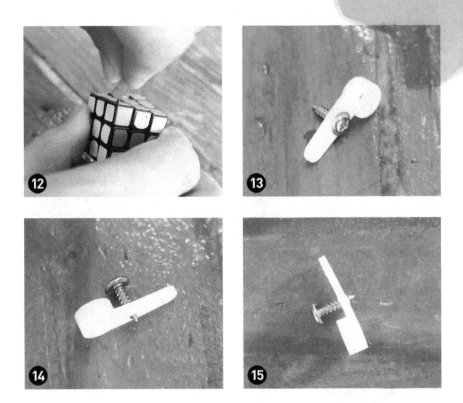

mixed side and at least one solved side (Figure **12**). Take your time with this to ensure that the hole you create is straight.

- Take out a one-armed servo horn and a mounting screw from your micro servo package.

- Mount the screw in the second hole from the center of the servo arm, screwing it in from the underside of the servo arm (the underside is the side that has the hole protruding.) We want the screw to be all the way in first (Figure **13**), and then unscrew it back out so only the very tip of the screw is peeking out on one side (Figures **14** and **15**). By doing this, we are pre-threading the hole for easier mounting later.

- Place the tip of the screw into the hole you created in the corner of the cube (Figure **16**).

- Hold the cube and servo horn in place as you slowly screw it in all the way until it is snug, but not too tight (Figure **17**). This is the most fragile part of the build! Once complete, set the cube and attached servo arm aside.

- Using your playing card box as a template, cut out seven pieces of cardboard (Figures **18** and **19**).

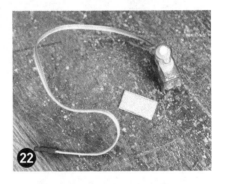

- Construct a 3D square with four pieces of cardboard, securing each side with hot glue (Figure **20**).

- Get out your micro servo. Using the bottom of the servo as a template, trace and cut out a small piece of cardboard (Figures **21** and **22**).

- Hot glue the small piece of cardboard to the base of your servo (Figure **23**).

- Press the servo horn that's attached to the cube into the micro servo without putting pressure on the cube itself (Figure ㉔).

- Hold the servo with the horn facing you, and gently turn the servo horn clockwise until it stops. We need it to stop at about the imaginary six o'clock mark... if it stops before or after, carefully lift the servo horn off and put it back on, so that the servo arm does line up properly (Figure ㉕).

- Adjust the cube by gently turning it so that the sides are aligned as in Figures ㉖ and ㉗.

- Attach the servo centered on the inside of one side of the cardboard square with just enough hot glue (Figure ㉘).

- Take another piece of cut-out cardboard from earlier, and cut it in half the long way (Figure ㉙).

- Use hot glue to attach one-half of the piece to the front bottom of your square structure (Figure ㉚).

- Cover the rest of the opening on that side of the square structure with another of the playing card-sized pieces of cardboard. Do not hot glue this piece in place, though (Figure ㉛).

- Use masking tape to attach this piece of cardboard to the lower piece, essentially creating a hinge (Figure ㉜). We want to leave an approximately ⅛" gap between the bottom cardboard piece's top edge and the top cardboard piece's bottom edge, bridged with masking tape. This will allow the front "door" to open easily.

- Take another micro servo, and with pliers, slowly bend and break off the mount on the side closest to the servo shaft. A few bends back and forth and a little twisting will do the trick. Set the servo aside while you complete the next step (Figures **33** and **34**).

- Cut a 5" piece of floral wire and thread it through the holes in a single-armed servo horn to create an arm extension, as you see in Figures **35**, **36**, and **37**. Use pliers to press it all snug. The whole servo arm with the extension should be 2¼" long.

- Attach the extended servo horn to the servo.

- Turn the servo horn clockwise until it stops. If it does not already land at the "12 o'clock" position, adjust the horn to the "12 o'clock" position by taking it off and putting it back on in proper alignment (Figure **38**).

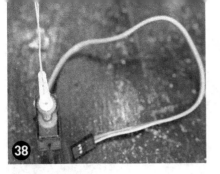

- Screw your servo horn in place (Figure **39**).

- Hot glue your servo to the right side of the inside lower front panel of your cardboard structure, sticking above the lower piece's top edge about ¼" (Figures **40**, **41**, and **42**).

- With the piece of cardboard that you cut the servo-sized piece from, cut one 2" × 1" strip and one 2.5" × 1" strip (Figure **43**).

- Place your last two Rubik's Cubes inside the

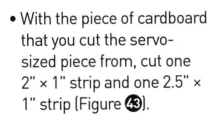

floor of your structure, on either side of the servo that's attached to the floor (Figure **44**). Set a small scrap bit of cardboard atop each cube, to add a little buffer for the next step (Figure **45**).

• Hot glue the 2" × 1" and 2.5" × 1" cardboard strips to the inner sides of the structure, as ceilings just a tad above each cube. The shorter strip will be on the side with the second servo attached (Figure **46**). You don't need a lot of glue... just a few dots along the edges will do. Be careful not to accidentally glue your cubes or scrap bits of cardboard! Remove the cubes and set them aside for later. Discard the scrap bits of cardboard (Figure **47**).

- Trim any excess cardboard that extends past the back of the structure.

- Using your playing card box as a template, cut a rectangular piece of dark felt (Figures **48** and **49**)

- Fold the felt piece in half widthwise as a guide, and cut halfway through (Figures **50** and **51**).

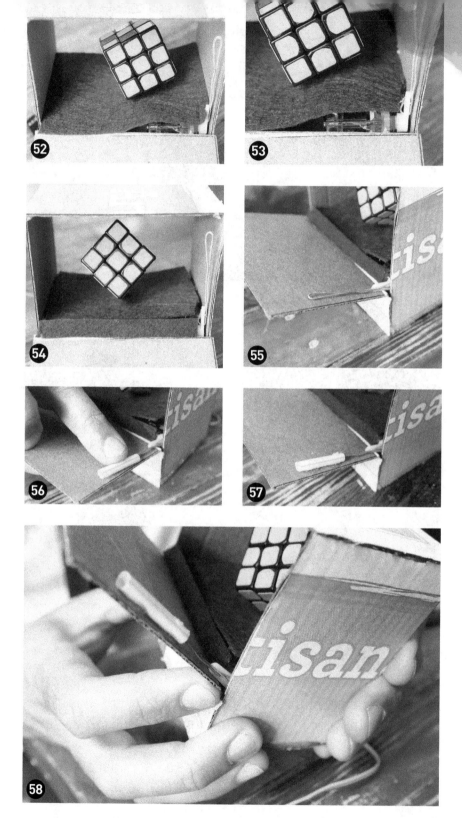

- Slide the felt piece into place from the front door opening of the square structure, so it acts as a "flooring" right under the servo-attached cube. The slit in the felt will allow it to fit around the servo shaft, but you may need to widen the slit slightly to avoid any obstruction of the servo movement (Figure **52**).

- Snip the right front corner of the felt a bit so the servo mounted on the side can move freely. You'll also want to adjust the felt along the edges for a good fit. If necessary, trim it slightly so it lays flatter on all sides. You may also wish to put a small loop or two of masking tape under the felt to hold it down (Figure **53**).

- Now cut an additional ¼" × 3.25" strip of dark felt, to cover the gap on the front edge of the felt. A very small amount of hot glue on each end will work perfectly to attach it in place (Figure **54**).

- The servo horn with the wire extension will make the structure's door open and close. Gently turn that servo horn — making sure to hold the servo horn and not the wire extension — so that the wire extension rides straight across the edge of the cardboard door (Figure **55**).

- Cut a 1" piece of plastic straw.

- Thread the straw piece onto the wire extension, covering half of the wire, from tip to mid-shaft (Figure **56**). Put a drop of hot glue to tack the straw down in position on the door (Figure **57**). Carefully close the cardboard door (Figure **58**) by pushing gently at the cardboard right outside the servo horn, making sure the wire does not bend.

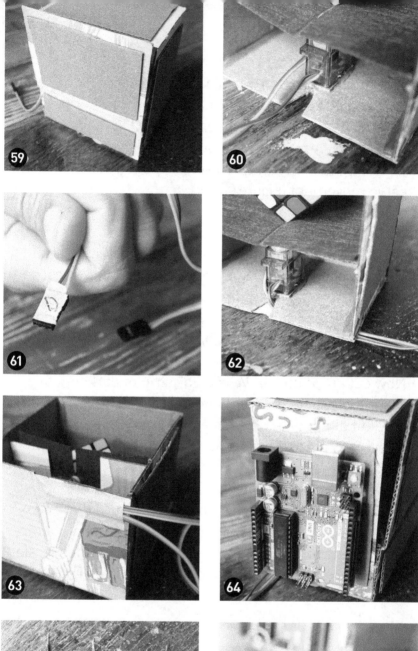

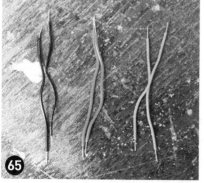

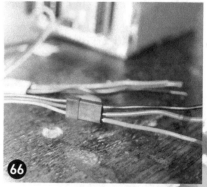

- If you'd like, for decoration, cut some smaller cardboard rectangles and mount them around the outside of the cardboard box structure with loops of masking tape (Figure **59**).

- Cut a slit out of the floor of the cardboard structure, from the center of the back edge just to the edge of the servo (Figure **60**).

- Place a small piece of masking tape to the end of the servo wires of the door servo. Mark it with a "D." This is so we will be able to easily identify it later (Figure **61**).

- Slide the wires from both servos through that slit, and tape them down on the bottom of the structure so the ends stick out on the side of the structure (Figures **62** and **63**).

- With a loop of duct tape, mount your Arduino board to that same outer side of the structure, with the USB port facing up (Figure **64**).

- Cut six 3" pieces of 22 gauge solid core wire: two red, two black, and two green. Strip the end of each wire about ¼" (Figure **65**).

- Connect your new wires to the servo wires: black to brown, red to red, and green to orange (Figure **66**).

- Now, let's connect the other ends of the red, black, and green wires to the Arduino board:
  - Both black wires go to GND on the board.
  - The door servo's red wire goes to 3v, and the cube servo's red wire goes to 5v.
  - The door servo's green wire goes to digital pin 9 on the board. The cube servo's green wire goes to digital pin 11 (Figures **67**, **68**, and **69**).

- Draw some eyes on an index card, cut them out, and tape them to the top of your cardboard structure. Our robot is starting to come to life (Figure **70**)!

- With a loop of duct tape, attach your 9v battery and plug to the cardboard behind the eyes (Figure **71**).

- Cut a thin strip of cardboard about 3" long and ³⁄₁₆" wide. Curl it into a ring and wrap it loosely around the base of the mounted cube, like a little collar. Carefully use a small dot of hot glue to tack it in place, without dripping any glue on the felt. The purpose of this is to hide the servo shaft and make it seem like the cube is just resting on this little cardboard base (Figure **72**).

During the next phase of this project, keep in mind that we'll need to test and adjust, test and adjust, tweaking angles, motor movements, and speed. The smallest of adjustments can take the illusion from okay to amazing!

Before uploading the sketch, gently unplug the two black GND wires from the board. Then, open the Arduino application on your computer and upload the sketch found at mariothemagician.com/robotmagicchapter7 to your board.

- Once uploaded, unplug the board from your computer, and plug both black GND wires back into the board.

- Plug the battery into the Arduino, and ten seconds will pass before the robot's mouth begins to open.

Now, let's go over the tweaks:

- The first thing to look for is your cube. Make sure the three mixed sides are facing forward and that none of the solved sides are visible from the front (Figure 73). If the cube is not positioned correctly, gently turn it into position.

- Make sure the door — the robot's mouth — closes and opens well with the code. In the closed position, the mouth should not be showing the cube at all. If it's not closing enough, gently bend the wire extension from the servo horn and adjust. If that doesn't do the trick, go back

73

a few steps and remount the servo horn in the correct position until it functions well.

The final time the robot's mouth opens, the cube should appear solved (Figure 74)! Once everything is working well, and the robot is successfully performing the cube illusion, we need to prep it for performance:

- With your two extra cubes, mix one up and keep one solved. Slide them both into their hiding places within the robot (Figure 75).

THE PERFORMANCE:

Alright! We have two loose hidden cubes and one mounted cube. Be aware that like most magic tricks, angles are important with this illusion! This works when your audience is directly in front of the robot.

• Hold the robot from the bottom with one hand (Figure **76**).

• With your other hand, push the battery plug into the board. This will start the routine!

• Immediately, tilt the robot back to remove the loose unsolved cube with your right cube, while holding the loose solved cube in place, so it doesn't fall out (Figure **77**).

• Show all sides of the unsolved cube to your audience (Figure **78**), and place it back in its hiding place. This all needs to be done within ten seconds, before the robot's mouth opens (Figures **79** and **80**). (Your audience will not

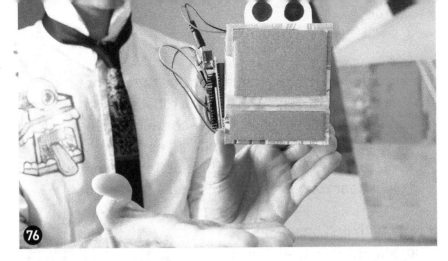

realize that the cube you just showed them is not the same cube that they see when the robot's mouth opens. To assist in this deception, keep your fingers close to the mounted cube before the mouth opens. As the mouth opens, pull your hand away. This subtlety strengthens the idea that you just placed that cube in the robot.)

- Now, with just one hand holding the robot, let it do its thing to reveal an instantly solved cube! BOOM! This is the big reveal! Ride this powerful moment (Figure 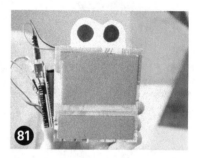)!

- Then, immediately after the robot closes its mouth after solving the cube, pinch the bottom of the robot by its sides. Use your fingers to block the loose mixed cube from falling allow the loose solved cube to fall into your hand, creating the illusion that you are removing the solved cube the audience just saw (Figure 82). Hold the loose solved cube up to show it on all sides to your audience as you unplug the robot and put it away (Figure 83).

83

Tips:

- In the code, just below **void loop()** you'll see **delay(10000);** — this is the 10 second wait period before the mouth starts opening. Feel free to change this number based on how you decide to perform the routine.

- Once the mouth opens you'll notice it moves at different speeds. We can control that in our code, too. Take the time to look over each line of code to understand how it works. Try changing the **delay(25);** to a higher or lower number and see how it affects the routine. For now, it's a slow opening and then a slow closing, then fast and slow again. I find that the small unexpected speed changes keep the audience alert, and when the surprise solve happens, it's even more striking. Play with your performance style and adjust the code to play along. See what works best for you!

#8

CLOWN NOSE EMOJI BOT

Two things I love smashing together are robotics and comedy. And one of the best ways to connect with younger audiences especially is a principle called "look, don't see." This is when something happens that the performer *seems* to be unaware of, but the audience can clearly see. This routine does just that.

THE EFFECT:

The premise is simple... we have a cardboard robot and a clown nose that just won't stay put. You try to attach the clown nose to the robot's face, but it won't stay attached and keeps falling off. As you reach down to pick up the fallen clown nose, another one appears behind the robot's head! You grab the nose from behind the robot's head, make it vanish, and it reappears behind the robot's head again! This repeats a couple of times, until, in frustration, you take the nose from behind the robot's head once again and turn it into a burst of confetti. You look back at the robot, and the clown nose has already magically re-appeared right where you always wanted it to be, smack dab into the middle of the robot's face.

Once built and performed, this routine will show you a glimpse of how I make kids laugh using robotics. We take traditional methods of children's magic that work and adapt the Arduino to do the heavy lifting. In terms of robotics, our goal here is to not only help you with programming multiple servos, but also to show you how to use a micro servo in a new way. Instead of only making something move, here we will be using a micro servo to release a reel. Remember, a reel can be attached to anything that it can pull, so there are a lot of possibilities that come along with this simple function.

Build this project and perform it. Study why it works in front of a live audience. Find a rhythm. Work on adapting your body movements and reactions to the robotics. Memorize your timing of when to take the clown nose and when to react to it appearing again. This is like karaoke, except that the "music" is made up of servo movements and clown noses! Once you've memorized this routine and are comfortable with it, take the foundation and make your own story with it!

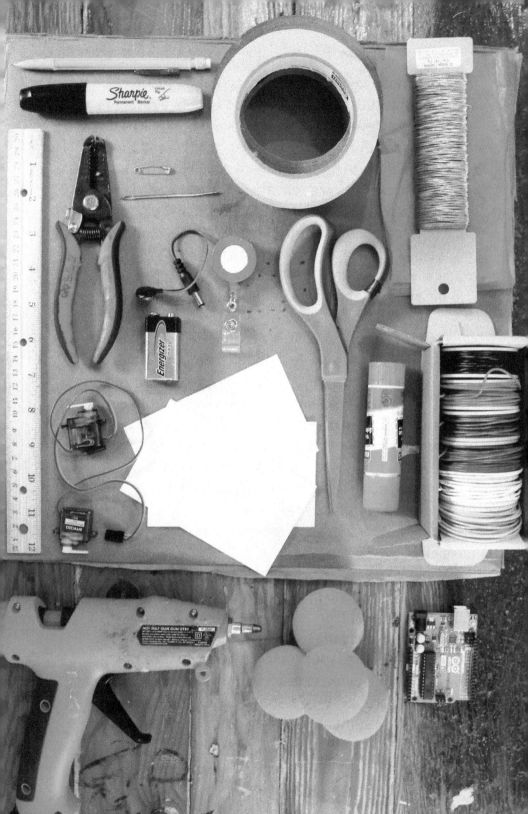

MATERIALS & TOOLS:

- 4 red sponge clown noses
- 1 name tag/badge holder reel with clip
  (you can find these at office supply stores)

 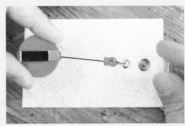

- 1 sheet of red tissue paper
  (or pre-cut red tissue confetti)
- 1 pizza box or cardboard of similar size
- 3 unlined index cards
- 1 small safety pin
- 2 micro servos
- 1 Arduino UNO
- 1 9V plug
- 1 9V battery
- 22 gauge floral wire
- 6 male-to-male wires
  (or 22 gauge solid core wire)
- Scissors
- Wire strippers
- Hot glue gun and glue sticks
- Duct tape
- Embroidery needle
- Masking tape
- Ruler
- Pencil
- Craft glue stick
- Black permanent marker

THE BUILD:

- Cut four 5.75" square pieces of cardboard (Figure ❶). Make sure each piece has at least one clean, unmarked side.

- Take one of the square pieces of cardboard and cut two 2.25" × 5.75" strips from it. These will be used as the sides for the robot face. Cut the remaining piece from that square in half, widthwise. These will be the ears! Set these pieces aside (Figures ❷, ❸, and ❹).

- Take two more of the square pieces, place them edge-to-edge, one above the other, and duct tape the seam together (Figure ❺).

- Tape the 2.25" strips of cardboard to either side of the upper square, with the strips positioned vertically, so the long ends are lined up along the sides of the square (Figure **6**).

- With a pencil, make a small mark 1" up from the bottom of each of the 2.25" strips, on the un-taped edge (Figure **7**).

- On both strips, use a ruler to help you draw a straight line diagonally from the pencil mark to the bottom corner of the taped edge of the strip (Figures **8**, and **9**).

- Cut the triangular piece out from each strip (Figures **10**, and **11**).

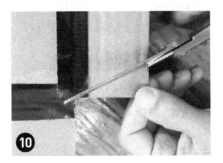

- Fold the lower square backward, and with duct tape, attach the inner edges of the diagonally cut strips to the inner edges of the lower square (Figure **12**).

- Take your last unused square piece of cardboard, and cut it in half, creating two rectangles (Figures **13**, and **14**).

- With a generous amount of craft glue stick, attach the two cardboard rectangles side by side to the outward face of the upper square of our structure, leaving a ⅛" wide gap in between the two pieces, running vertically (Figures **15**, **16**, and **17**). Make sure all edges are pressed and glued very well. This front-facing area will be the face of our robot.

- With your black permanent marker, carefully fill the vertical gap between the two pieces of cardboard with a dark black line (Figures **18**, and **19**).

- Measure 2½" up the black line from the bottom of the robot's face, and create a hole with your embroidery needle, right through the black line (Figure **20**).

- Take your name badge reel, and pull the string out about 6" or so (Figure **21**).

- Fold a piece of masking tape onto the string at the base of the reel, so it won't retract back into the reel as we're working with it (Figure **22**).

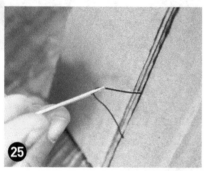

- Snip the plastic attachment off of the end of the reel string (Figures **23**, and **24**).

- From the backside of the robot's face, push the end of the reel string through the hole we made (Figures **25**, and **26**). (If you have trouble with this, you may find it easier to thread the reel string through an embroidery needle first.)

- Tie a small safety pin to the end of the reel string. A good double knot will do (Figure **27**).

- Take one of your sponge clown noses, and attach the safety pin with reel

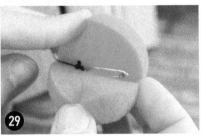

string right inside the center of the nose's opening/crease (Figures 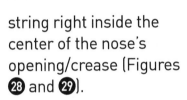 **28** and **29**).

- Completely remove the piece of masking tape from the reel string.

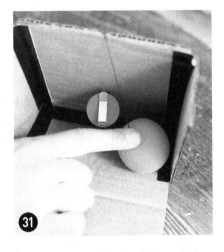

- Let's take a moment to test our clown nose reel! Gently pull the clown nose up over and behind the cardboard face — you'll now see the purpose of the black line, as it camouflages our reel string (Figures **30** and **31**). Then, let it go. The nose should spring right back onto the face (Figure **32**). (If it doesn't, adjust the reel and tape so the string is pulling and retracting

33

smoothly, and try again. If the string is catching on the cardboard face plates, you may need to re-glue the edges, so they are extra secure.)

- Next, let's frame the robot face with masking tape to clean up and secure all edges (Figure **33**). Take your time with this! The neater these details are, the cleaner the whole routine will look.

- With your marker, redraw the black line where the masking tape covered it on the top and bottom of the face (Figure **34**).

- Take the last two small cardboard pieces we had set aside, and tape them to the sides of your robot head, as robot ears (Figure **35**)!

34

35

- Now that we've completed our robot head structure, it's time to make the robot's eyes and mouth! Cut two 2" circles and one 3"×1" rectangle from index cards, and outline them all in black marker. Draw pupils at the center of each circle and teeth lines in the rectangle (Figure **36**).

Attach them to the robot face with loops of masking tape, making sure not to block the black line with the eyes. The mouth isn't a concern, as long as it's positioned below the puncture point in the black line (Figure **37**). Our robot face is complete!

Now, we can move onto the electronics! Take out a micro servo, duct tape, a clown nose, floral wire, and wire strippers.

- Turn your robot face around, so the back side is facing you. Use a small loop of duct tape to secure the servo to the top left corner of the face plate, with the servo shaft pointing to the right, and the wires to the left. Leave about ½" of space from the top edge and side (Figures **38** and **39**).

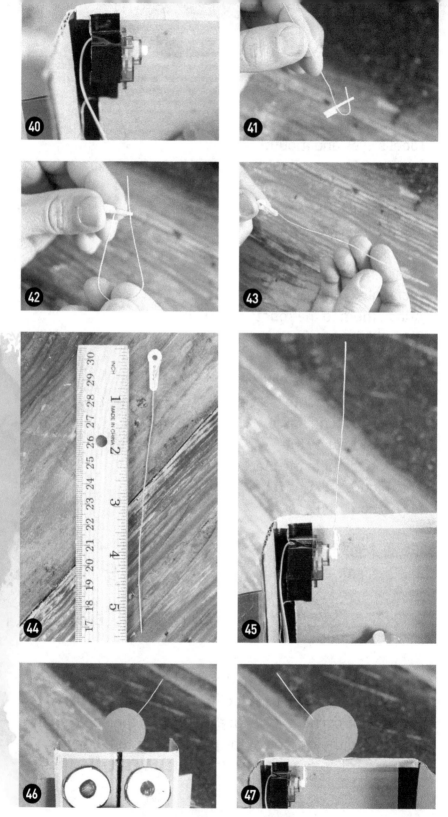

- Add a thin strip or two of duct tape over the servo for extra secureness (Figure **40**).

- Cut an approximately 6" piece of floral wire. Then thread, bend and wrap it onto a one-armed servo horn to create a 5.5" servo arm extension (Figures **41**, **42**, **43**, and **44**).

- Press the wire-extended servo arm onto the servo shaft so the wire is sticking straight up (Figure **45**).

- Poke the end of the wire through the split area of the clown nose and slide the nose down until it is low enough on the wire to just clear the top of the robot face. Adjust and bend the wire, so that in up-position, the clown nose is sitting just above the center of the robot face, without any wire visible from the front (Figures **46**, and **47**).

- Once in position, twist the remaining wire back around itself and make final adjustments to the clown nose position (Figures **48**, and **49**). Being careful not to bend the wire out of place, pop the servo arm off the servo, and set the arm/wire/nose piece aside for now.

- Now, let's work on the electronics for the clown nose on the reel! Use hot glue or duct tape to secure a second micro servo to the floor of the inner base, 2¾" in from

the right and back edges, with the servo shaft pointing in toward the face of the robot (Figure **50**).

- Attach a one-armed servo horn to the servo, with the arm pointing to the right, almost touching the floor of the structure (Figure **51**).

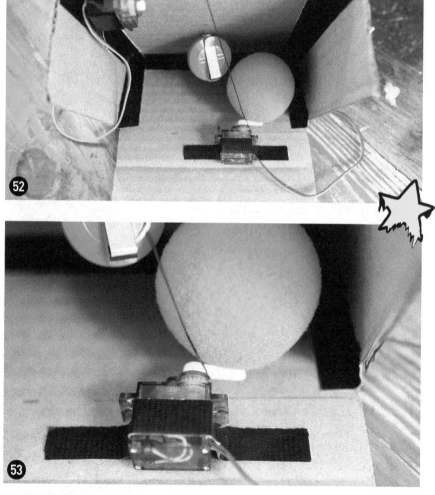

- Gently pull the clown nose on the reel out and over the robot face, following the black line. Loop the string under the servo arm, so the ball rests snugly in place (Figures **52**, and **53**). This will all hold together until the servo arm moves up! At that point, the nose will fly out and stick right onto the robot's face! If you can reach, manually move the servo arm up (or just slide the string off) to see how it works. Pop the servo arm off carefully, and set it aside for now. We'll get back to our servo arms later.

- Make a loop of duct tape, and secure your Arduino board to the floor of the cardboard base, right at the bottom left corner, with the USB port facing back toward you (Figure **54**).

- Now, let's connect our servos to the Arduino with male-to-male wires (Figure **55**). We'll begin with the servo that's taped to the floor of the cardboard base. Connect the orange/yellow wire from the servo to digital pin 9 on the Arduino. That's your signal. The red wire from the servo gets connected to the 3.3V pin, and the black wire from the servo goes to any one

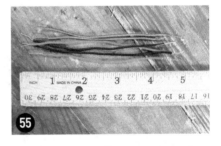

of the three GND pins (Figures **56** on the previous page, **57**, and **58**).

• Moving on to the top-mounted servo, use male-to-male wires to connect orange/yellow to pin 8, red to 5V, and black to GND (Figures **59**, **60**, and **61**).

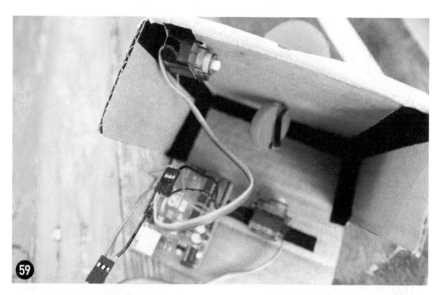

- Open up your Arduino application from your computer. Copy and upload the sketch found at mariothemagician.com/robotmagicchapter8 to your board.

- Listen for the servomotors after the sketch is uploaded. They will both turn into starting positions. If you do not hear any motor sounds after upload, go back and check that all wire connections are in the proper position and secure. After the sketch is uploaded and everything is connected properly, let the board stay on for about five seconds before pulling out the USB cord to turn it off.

- Once the Arduino is off, gently and carefully bend and tuck your wires out of the way, careful not to disconnect any, and re-connect the servo arms each to their respective servos. The servo arm with the clown nose attached should be positioned pointing downward, hidden behind the robot face. Pull the clown nose on the reel into ready position, as earlier described (Figure **62**).

- Let's test and make minor adjustments! Plug in the Arduino, wait for the clown nose on the wire to pop up. (This will happen ten seconds after turning on the Arduino.) Once it pops up, quickly unplug the board, so the clown nose stays in up position. Now, you can look at the robot from the front and carefully bend and adjust the wire if needed, so that the nose is clearly visible, centered, and not showing wire from the front. Once it looks good, plug the board back in. The nose will move back and hide from view, and you are all set!

NOTE: if the servo horn happens to pop off as you are making adjustments, don't worry! Simply plug the Arduino back in, and allow the servo to move back to its hiding position, then unplug the Arduino, and pop the servo horn back on.

- With a loop of duct or masking tape, secure a 9V battery to the right outer corner of the cardboard base. Attach the battery plug to the battery but not yet to the Arduino (Figure **63**).

- Cut some red tissue paper into roughly 1" squares to make confetti. A small stack is plenty. Place the confetti in a neat little stack on top of the 9V battery (Figure **64**).

- Take an extra clown nose, and rest it on top of the Arduino (Figure **65**).

And now, we're officially ready to learn the performance!

63

64

65

THE PERFORMANCE:

- Stand facing the audience with the robot to your left and positioned ½" from the front edge of the performance table (Figure 66).

- Plug in the Arduino, take out the loose clown nose, place it on your own nose, smile and bow to your audience (Figure 67).

- Point to your nose and then to the robot's. Slowly take the clown nose off and start trying to attach the nose to the robot's face. This is all pretend, of course. It won't stay. After two or three attempts, allow the nose to "accidentally" fall to the floor on your final failed attempt.

- Bend down to pick up the nose, and as you stand up, the nose you had just picked up vanishes from your hand* as another nose appears on the robot's head.

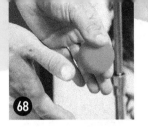

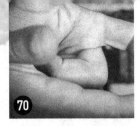

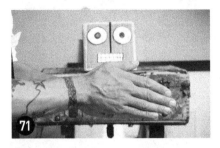
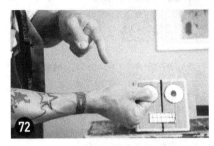

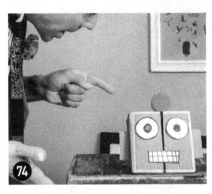

*This vanish is created with a sleight of hand move called a false transfer. Start by turning your body toward the robot, with your right side toward your audience. Bend down to pick up the nose with your left hand's index finger and thumb (Figure 68). As you stand back up, bring your hands together, seemingly to transfer the nose from your left to right hand. As your hands touch, almost like a clap, at that very moment, squeeze the nose into a finger palm position with your left hand (Figures 69 and 70). Be mindful to keep your right hand in position to shield this movement from the audience. This is where your middle, ring, and pinky fingers curl in, hiding the nose as your index finger stays extended. As that happens, you immediately close your right hand, as if it received the ball (Figures 71 and 72). You end up with your left hand pointing toward your right hand, which has a closed fist. Open your right hand to find that the clown nose is gone (Figure 73)! Act surprised as you also notice that it's on top of the robot (Figure 74)!

- The ball on top of the robot hides itself again, so you reach behind the robot to grab it, but rather than actually grabbing that ball, you use your left hand that's still in finger palm position as you reach behind the robot's face, and bring out that ball that you had actually been hiding in your hand (Figures 75 and 76). Make sure to not flash (accidentally show the audience) the finger palmed nose too early! Practice the motion in front of a mirror to get the timing right.

- Now, perform the false transfer again. Once the finger is hidden in your left hand, and your left hand is pointing toward your right closed fist, puff your cheeks up with air, and make a long exaggerated blowing sound towards your right closed fist, as you slowly open up your hand (Figures 77 and 78). Of course, the nose is gone again, and at the same time, a nose appears once again atop the robot (Figures 79 and 80)!

- As the nose from the robot hides again, repeat the process. Pull the finger palmed nose from behind the robot's head, then false transfer yet again. This time, instead of blowing into your hand to reveal the vanish, take your closed right hand, and pretend to eat it! Your finger palmed left hand points to your mouth as you chew. The nose again appears for the third time back atop the robot as you are chewing (Figures **81**, **82**, and **83**).

- Again, reach behind the robot's head to grab the nose, and pull out the finger palmed nose (Figure **84**), but instead of vanishing it yet again, simply stuff the nose into your front pocket in frustration (Figure **85**). Leave the

nose in your pocket, and pull your left hand out. Swipe your palms clean, as if it's all done (Figure **86**). As you do so, the nose pops up atop the robot one more time (Figure **87**).

- Reach behind the robot again, this time grabbing the stack of confetti. Quickly flashing red as you bring both hands together, cup your hands with the confetti inside. Quickly stand in front of the robot, and blow the confetti toward the audience (Figures **88** and **89**)!

- As you step away, you reveal that the clown nose finally appears where it belongs, right on the robot's face, and you can take your bow (Figure **90**)!

This is a routine that will definitely take some practice! Take your time to practice the timing of each move. Mirrors help a ton! It's a series of quick motions and movements. After a few tries, it will all sink right in, and you'll learn the pace and rhythm. I cherish this routine so much, because it's near and dear to my personal performance style. You can also go through the code line by line and make changes to it to suit your own pace and style! Slow it down, if you'd like. Change the delay portions, and have the nose appear between longer wait periods. Each line in the sketch is marked with notes for you, so you can hack it and make it your own.

#9

RAY GUN CONTROL STATION

THE EFFECT:

We see a miniature cardboard control board, a ray gun, and an inflated clear balloon with a smaller inflated black balloon inside of it. The performer explains that within this balloon lies a prediction of future events to come. They show that the control board has five different colored LEDs and a single button. When the button is pressed, all the LEDs flash on and off randomly. When the button is pressed again, just one LED remains on. Each time the button is pressed on and off, a different colored LED will remain on, in no particular pattern. An audience member is asked to press the control board button on and off, randomly selecting a color.

The performer now presents the cardboard ray gun, explaining the dangers of lasers and how they can create intense heat and can even cause blindness when used without care. They explain that lasers are also able to pop balloons, but because black balloons absorb light and heat more quickly, if the laser is pointed at the black balloon, only the black balloon will pop, leaving the clear balloon intact. The performer reminds everyone that the balloon has been visible the entire time, well before any color was selected with the control board. The audience member is asked to concentrate on their randomly chosen color and aim the laser toward the balloon. Within a few moments, the black balloon pops, revealing a burst of confetti inside the inflated clear balloon that matches the randomly chosen color.

THE BUILD:

- Cut two 2.5" × 3.75" rectangles out of cardboard. (Alternatively, you can trace a standard playing card box as a template.) Also cut two 1.25" × 2.5" pieces and one 1.5" × 3" piece (Figure ❶).

- With your pencil, draw out a retro ray gun shape from cardboard. Cut it out, and outline it with your marker (Figure ❷). The ray gun shape should be no larger than 5.5" × 4.5."

MATERIALS & TOOLS:

- 5 LEDs (all different colors, including red), 5mm diameter
- 5 270 Ω resistors
- 2 momentary buttons
- 1 10K resistor
- 1 3-5V mini laser head diode module OR a common laser pointer
- 1 6V button coin cell battery holder (to power laser head module)
- 2 CR2032 coin cell batteries
- 1 Arduino UNO
- 1 USB 2.0 cable
- 1 9V battery plug
- 1 funnel
- 1 white or yellow round head sewing pin
- 22 gauge solid core wire (black, green, red)
- 1 spool 0.6mm solder wire
- 1 piece of cardboard
- 1 bag of 12" clear balloons
- 1 bag of 9" opaque black balloons
- 1 sheet of red tissue paper
- Hot glue gun
- Hot glue
- Soldering iron
- Wire strippers
- Wire cutters
- Scissors
- Small screwdriver or embroidery needle
- Ruler
- Masking tape
- Duct tape
- Pencil
- Black permanent marker

- Take one of the 1.25" × 2.5" pieces, and hot glue it standing up perpendicular to one of the shorter sides of a 2.5" × 3.75" piece. The larger piece is the base of the control board, and the smaller piece standing up is one of its sides (Figure ❸).

- Make a small cutout, 2" inch long and ½" deep, in the center of one of the longer sides of the other 1.25" × 2.5" piece of cardboard. This cutout will give us access to the Arduino later (Figure ❹).

- Glue this piece standing up on the other short end of the base, with the cutout facing down. Reinforce the inside edges of both sides with extra hot glue, if needed for stability (Figure ❺)

- Mount your Arduino UNO onto the base with a loop of duct tape. It should be positioned so that both the USB and 9V plug are easily accessible through the side cutout (Figure ❻).

- With your marker, draw a border with a ¼" margin on your 1.5" × 3" piece of cardboard (Figure ❼).

- With your pencil, evenly space out five dots lined up within the border. This is where the five LEDs will go (Figure ❽).

- With one blade of your scissors, carefully poke a hole through the cardboard at each dot (Figure ❾).

- Each LED has a round top and a small, flat-edged base. From the back of the cardboard, carefully twist and push one LED (any color except for red) head first through the

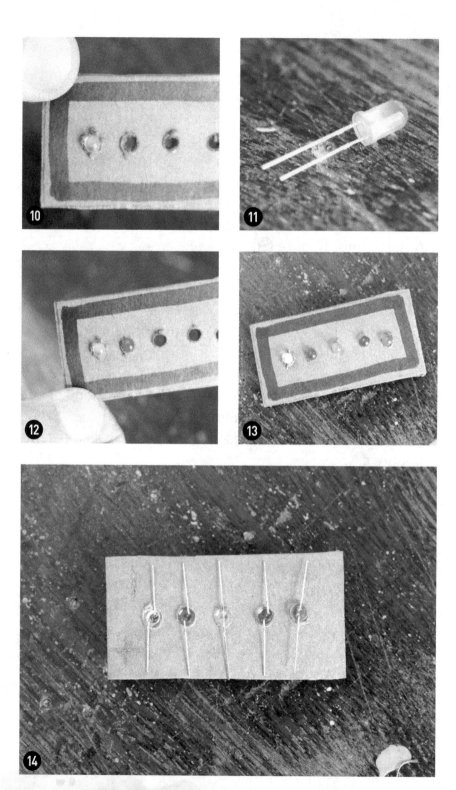

first hole in the row, so that the round top is showing on the front side and the edge is flush to the back of the cardboard (Figure ❿).

Before inserting the next LED, we have a few very important things to pay attention to in order for the whole project to work smoothly:

- First, notice how each LED has two wires, one slightly shorter than the other (Figure ⓫). The shorter wire is ground, and the longer wire is signal, or positive. Keep in mind that when you press each LED in, you will need to separate the wires so that all the shorter wires are on one side and all longer wires are on the other side. With your pencil, make a " – " (negative) mark on the side for the shorter wires and make a "+" on the side for longer wires. This will help us a lot later! Also note that when bending our wires, do so carefully and sparingly. Each time we bend a wire it becomes weaker, and these very small wires can break pretty easily. But don't insert the LEDs yet, because...

- The last important thing to note before moving forward is to make sure that the second LED from the left, when looking from the front, is a *red* LED (Figure ⓬)! This correct positioning will eliminate a lot of headaches moving forward. Once we are certain the second LED is red, and our wires are lined up with all ground wires on one side and all positive wires on the other, we can continue (Figures ⓭ and ⓮).

- It's time to solder! Turn your soldering iron on. The ground wires, on the " – " side, can be soldered together to one single wire. To do so, first, tape down your

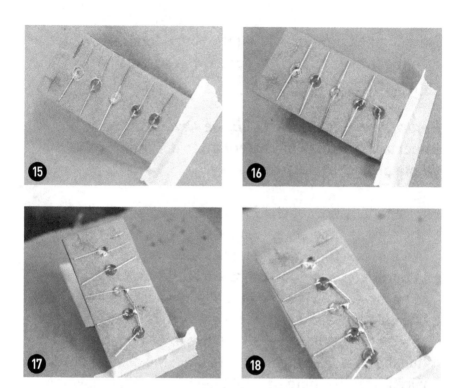

cardboard with masking tape, so it doesn't move
(Figure **15**).

- Start by bending just one wire so its end is touching
  the base of the next wire (Figure **16**).

- Use your soldering iron to connect the first wire to the
  base of the second wire. It will only take a little heat
  and solder to accomplish that.

- Once your first two ground wires are soldered together,
  carefully bend the end of the second wire to the base
  of the third ground wire, and solder those together
  (Figure **17**).

- Repeat, connecting the end of the third ground wire
  to the base of the fourth (Figure **18**).

NEVER SOLDERED BEFORE?

Don't be deterred! Here's a simple method. Once your soldering iron has heated up, carefully hold it like you would hold a pen and hold the spool of solder in your other hand, with a few inches of it unspooled (Figure **A**). Carefully position the tip of the iron so that both of the first two ground wires are touching it (Figure **B**). Wait about one second, then slowly press the solder to the point where the tip of the iron meets the wires (Figure **C**). The solder will liquify both wires together. Once you see that happen, move the iron and solder spool away (Figure **D**). This should not take more than a few seconds, and you'll only need a very small amount of solder. Practice with some 22 gauge wire first, to help give you a feel for it before working with the LEDs. It's also good to have a soldering iron tip cleaner on hand (Figure **E**). A few brushes after each time you solder will help make the solder stick better and extend the life of your tool.

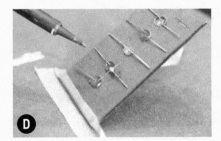

- Finally, connect the end of the fourth and fifth wires together (Figure ⓳).

- Next, cut a 4.5" piece of 22 gauge wire, and strip the coating off roughly ¼" from each end. (If you have a black coated wire, use that for this step! Whenever possible, I try to consistently use black for ground, green for signal, and red for positive.)

- Now, solder this wire to the ground wires at the point where the fourth and fifth wires were connected. Release and let cool (Figures ⓴ and ㉑).

- The next step is to solder a 270 Ω resistor to each of the LED's positive wires. If you are familiar and comfortable with soldering, by all means, proceed! If you are new to soldering, here is a simple method:

  - Start by clipping the positive wires from the LEDs so they are each about ¼" long (Figure ㉒).

  - Place a small amount of solder on each end (Figure ㉓).

  - Take out five 270 Ω resistors, and clip one end from each to about ¼" long (Figure ㉔).

  - Take one of the resistors, and thread the short end under one of the positive LED wires, bending it over, so it temporarily stays in place (Figure ㉕).

  - Solder it on (Figure ㉖)!

  - Do the same exact thing with the other four LEDs. Take your time with this, making sure the resistors are

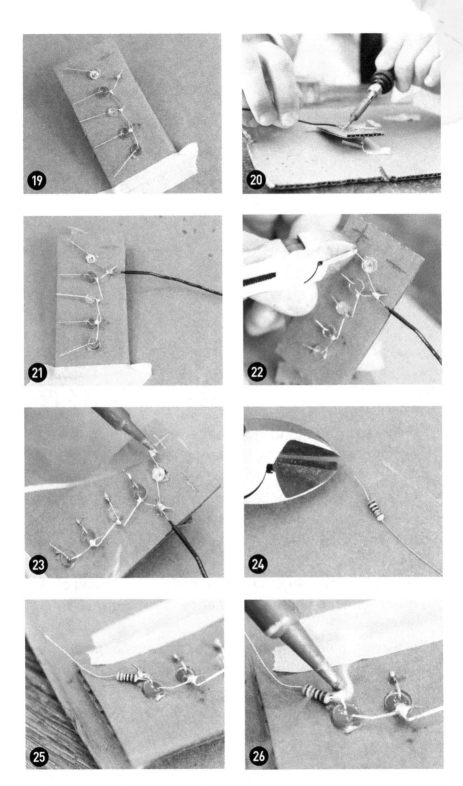

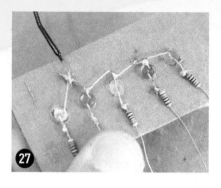

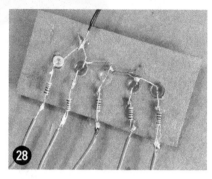

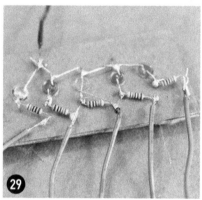

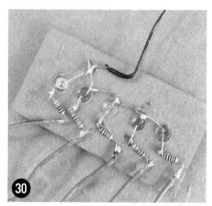

soldered on straight. The cleaner everything looks, the easier it will be moving forward (Figure **27**).

- Cut five 4.5" pieces of 22 gauge solid core wire, and strip the ends at around ¼". If you have green coated ones, use those for this step!

- Solder each long end from the resistors to the end of a green wire, one green wire per resistor (Figure **28**).

- Once everything has cooled down, carefully bend all the green wires so that the exposed metal ends are within the edges of the cardboard (Figure **29**).

- Do the same thing with black ground wire and any other long wire tails that could possibly touch other wires

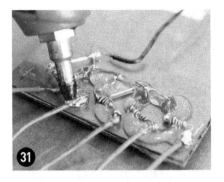

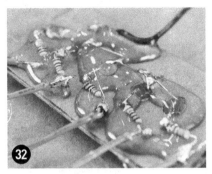

accidentally (Figure **30**).

- Make sure no wires are touching each other, then put a thin layer of hot glue over everything on this side of the cardboard, especially the areas where the black and green wires are attached to the LED wires. The hot glue will not only seal and protect everything, it will also prevent these thin wires from breaking as we begin to mount everything together (Figures **31** and **32**).

- Attach the loose end of your black (ground) wire into the Arduino board's GND pin — the one directly after digital pin 13 (Figures **33** and **34**).

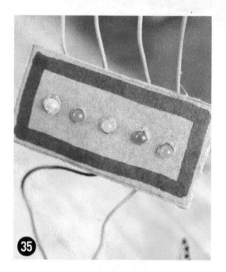

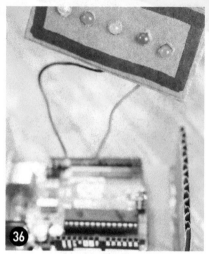

- Let's get the signal wires attached to our Arduino! When facing the LEDs, start with the far right LED's green wire, and push that into digital pin 8 on the Arduino, carefully bending the wire downward (Figures **35**, **36**, and **37**).

- Moving down the line from right to left, push the next signal wire into digital pin 9, the following to pin 10, then pin 11, and the last wire on the far right goes to digital pin 12 (Figures **38**, **39**, and **40**). Double check that the red

LED is connected to digital pin 11. This is vital.

- Take your last 2.5" × 3.75" piece of cardboard, and with your marker, draw a rectangular border with a ¼" margin (Figure **41**).

- Draw a 1" square toward the bottom left corner within the border (Figure **42**).

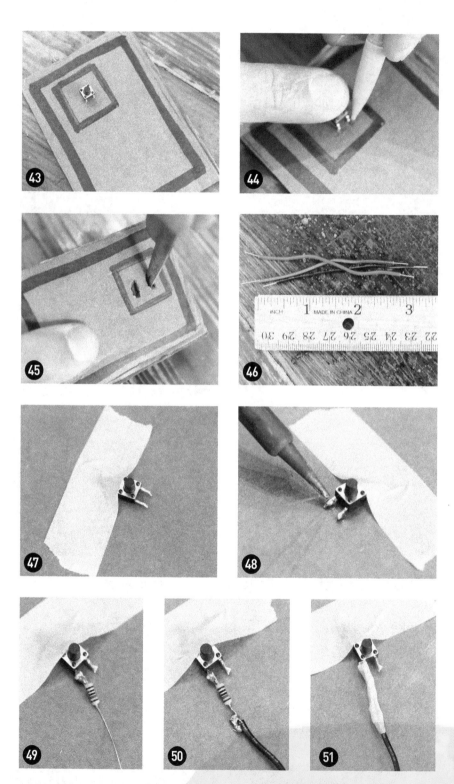

- Temporarily rest your momentary button in the center of the 1" square, and with your pencil, draw a little line to the left and right of the button (Figures **43** and **44**). Set the button aside.

- With one blade of your scissors, carefully make a slit at each pencil mark (Figure **45**).

- Cut a black, green, and red wire at 3" each, and strip them around ¼" on each end (Figure **46**).

- Tape a scrap piece of cardboard onto your work table, and place the momentary button on it. Carefully bend the button wires out a little bit, and tape down one side of the button with a piece of masking tape (Figure **47**).

- Put one small drop of solder on each of the two wires on the other side of the button (Figure **48**).

- Take one end of your 10K resistor and wrap it around one of the soldered pieces, then solder it on (Figure **49**).

- Wrap the other end of the resistor to the end of the 3" black wire, and solder it on (Figure **50**).

- The wire from the resistor is very fragile! If you have shrink tubing available, adding that to the connection would be ideal. You can also use a thin layer of hot glue along the soldered points and wire. Even a tight, small, rolled piece of masking tape will do, This will help strengthen the wire so it can be gently bent without breaking later on (Figure **51**).

- Locate the wire on the button that is directly across the

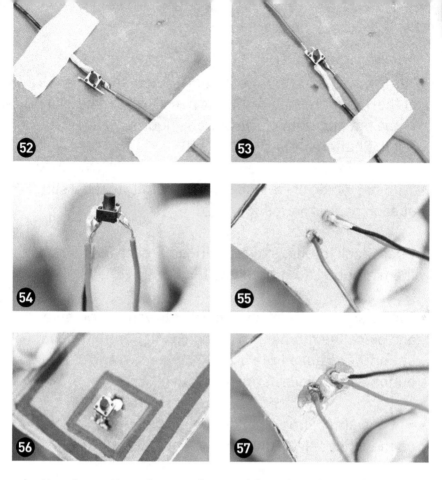

button from the wire you just soldered to the resistor. That wire needs to be un-taped and soldered to the 3" green wire (Figure **52**).

• Solder your 3" red wire to the other wire on the button that already has solder on it — the one next to the resistor (Figure **53**).

• Once all three wires are soldered, carefully bend each wire so they point downward (Figure **54**).

• Now it's time to mount the button in position inside the 1" square on the cardboard panel. Slide the button wires through the slits we created earlier, but don't force them! If they do not slide through easily, widen the holes,

then try again. It's much easier to widen the holes than to re-solder the button. And remember, the wires on both button and resistor are very fragile, so work gently (Figures **55** and **56**).

- Once mounted, add a few drops of hot glue on the underside of the cardboard, right where the wires come out. Let the hot glue dry completely before moving on (Figure **57**).

- Once completely dried, carefully plug the green wire to digital pin 5 on your Arduino board, the red wire to 5V, and the black wire to a GND pin (Figures **58**, **59**, and **60**).

- It's programming time! Open the Arduino software on

your computer, and copy and upload the sketch found at mariothemagician.com/robotmagicchapter9.

- Let's test it! Once the sketch is uploaded, press the button! All the lights should flash randomly and continue flashing until you press the button again. Then, all the LEDs should stop blinking except for one random LED, which stays on. Repeat and try again! We've created a random LED generator! If something is not lighting up or the button is not working, go back, check all the wires, and make sure they are all connected well and in the correct position. Also, make sure all the wires soldered to the button are still connected.

Once everything is in working order, let's take a quick look at the code. In the sketch, you'll see the "**//**" notes before every major action. Read those lines to understand what's happening within each section. The LEDs are all mounted in a line on the cardboard panel. In the code, you'll see the LEDs labeled from left to right **ONE** to **FIVE**. This is so you can clearly follow along in the sketch to understand which LED is turning on at which time. Now, with each button press, we activate a new "case." Moving from **case 1:** to **case 10:** Each case is performed in order and is activated and changes when the button is pressed. Within each "case" are a series of actions. In this particular example we go between two major actions, one being a series of random blinking LEDS and the other being a single LED that stays on. This changes slightly as we go from case to case. Again, this is a random LED generator... or... is it?

The secret to this routine? You may have already noticed from reading the sketch that there is *actually no randomness involved at all!* Press reset on the Arduino and let's do it again. The first press of the button makes the LEDs flash in a seemingly random order. The truth is, under **case 1:** you can see the LEDs blinking in a very *specific* order. Press the button a second time and one seemingly random LED stays on. In actuality, it's not random at all! It is all pre-programmed specifically to seem random to an audience, while the performer remains in total control of the outcome.

But before we get to the performance technique, let's finish our build:

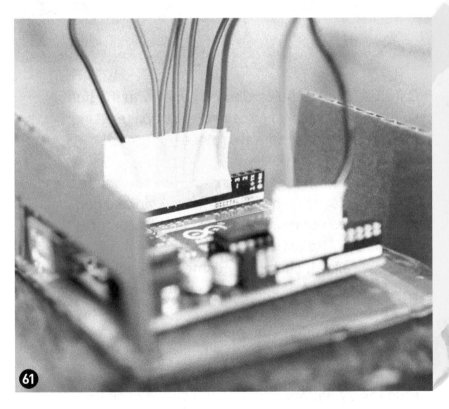

- Let's use masking tape to further secure the wires that are attached to the Arduino board. This will help prevent any pulling in our next steps (Figure **61**).

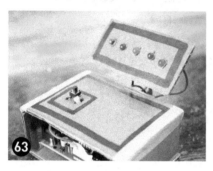
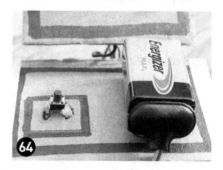

- Carefully mount the cardboard panel with the button, so that it acts as a lid or ceiling to our base, using masking tape to secure it in place (Figure **62**).

- Next, carefully bend the signal wires attached to the LED panel, so that the LED panel sits angled and visible above the rest of the cardboard structure (Figure **63**). After, check the wires below to make sure they are all still plugged into the headers and none are overlapping with each other.

- With a loop of duct tape, mount the 9V battery to the right of the button on the cardboard panel (Figure **64**).

- Plug the 9V battery plug into the battery and the Arduino, and do a final test to make sure our LED display is working! Once satisfied, unplug the battery from the board, and set it all aside.

RAY GUN BUILD:

If you already have a laser pointer, you can skip this step, and instead just tape your laser pointer to your ray gun cardboard cutout! If not, or if you're just up for a little more tinkering, let's make one from scratch! Take out your cardboard ray gun, another momentary button, the laser diode module, a coin cell battery holder, and coin cell batteries.

- Start by soldering the red wire from the laser diode to one of the wires from the button.

- Solder the red wire from the battery holder to the wire closest to the one you just soldered on the button.

- Solder the black wire from the battery holder to the remaining wire from the laser (Figure **65**).

- Put the batteries inside the holder, point the laser away from all living eyeballs, and press the button. When all connections are sound, the laser will light up (Figure **66**)!

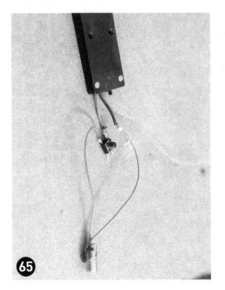

65

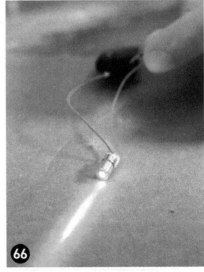

66

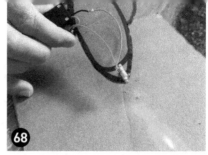

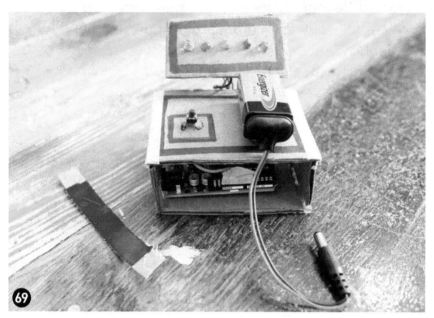

- Use a loop of duct tape to mount the battery holder onto your ray gun, and with a drop of hot glue, mount your laser diode to the front end of the cardboard ray gun. With another drop of hot glue, mount the button into "trigger position" on the ray gun handle (Figure **67**).

- After the glue is completely dried, test it all once again before setting it aside (Figure **68**).

- Make a simple holster for the ray gun with some duct tape folded and mounted on one side of the control board (Figures **69** and **70**).

- Slide your ray gun in place, and your control station is complete (Figure **71**)!

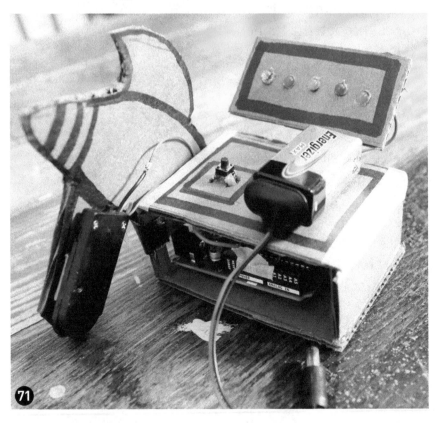

BALLOON PREP:

- Let's get to the balloon prep! Blow up and deflate one black balloon and one clear balloon. This will help later on.

- Create red confetti by cutting a handful of ¼" squares of red tissue paper (Figure **72**).

- Using the funnel, load the black balloon with the confetti, using a pencil to gently push the confetti down through the funnel (Figure **73**).

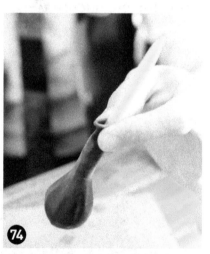

- Remove the funnel from the balloon, and stick your pencil halfway into the black balloon nozzle, eraser end first (Figure **74**)!

- Next, we have to insert the black balloon into the clear one, slowly, and with the assistance of the pencil. Take your time (Figure **75**)!

- Inflate the clear balloon! Start by stretching the mouth of the clear balloon

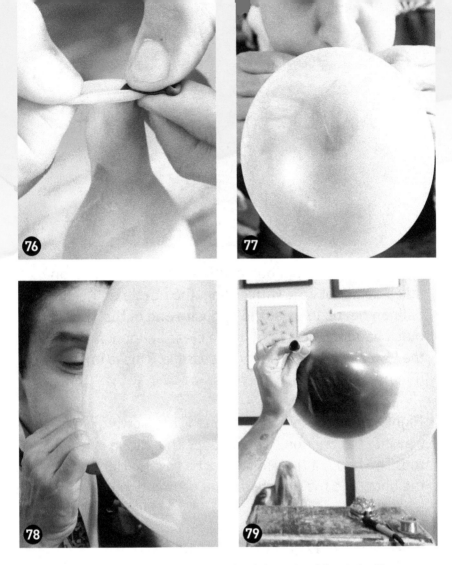

between both hands, while pinching the black balloon to the side. Fully inflate the clear balloon (Figures **76**, **77**, and **78**).

- Pinch everything closed, and now inflate the black balloon so that it takes up about 60% of the clear one (Figure **79**).

- Tie a knot in the black balloon and push it all the way in, so that the black balloon is moving freely inside the clear balloon.

- Deflate the clear balloon a little, so the black balloon takes up about 75% within. And, the nozzle of the clear balloon needs to be at the opposite end of the knot of the black balloon. Tie a knot in the clear balloon, *but make the knot as close to the end as possible*. This is important (Figure 80)!

- Right above the knot, slowly push your sewing pin all the way into the balloon. It won't pop! That alone is magic to me! The head of the pin should be resting right next to the knot, and the needle should be pointing inside the balloon (Figure 81).

We are now ready for performance!

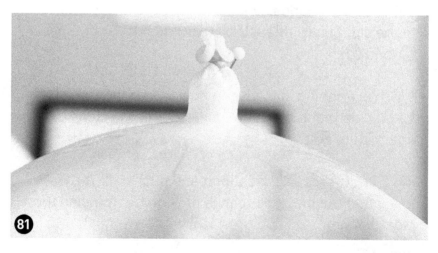

THE PERFORMANCE:

You already know that the LED random color generator is not actually random. You also know by now that the red LED — the second LED on the panel — is the one we need our spectator to choose. But how do we take something predetermined, and make it *feel* completely random to our audience? Let's get into it!

- Turn on the Arduino, and introduce the control board to your audience, explaining, "This is a random LED color generator that I built! Look! When I press the button, all the LEDs flash randomly. When I press it again, it stops on one of the LEDs. See? We got blue this time." (Or whatever color you have at position #4.) "Look! I press it again, and it goes back to flashing. It will continue flashing until the button is pressed again. Look now it's landed on red! Here! You give it a try!" Hand the generator to the spectator. The spectator presses the button and all the LEDs flash again. You say, "Anytime you want to press the button again, go ahead!" They press it, and you say, "Look! It's green!" (Or whatever color you have at position #5.) You say, "Okay... now I'm going to turn away, and I want YOU to randomize the LEDs, then press the button whenever YOU want and memorize the color you stop at. Got it? Good! Remember that color, then press the button again, so the lights are all flashing again." (We have programmed the LEDs, so the fourth time the sequence is run, it will always be red.) Ask them to concentrate on their color. Remind them they could have stopped at any color. Press the button again, so it stops again at a different color, as one final convincing measure of randomness. Unplug the board, put the generator away, and grab the ray gun!

- Introduce your homemade laser ray gun. Explain how dangerous lasers are, how they can cause instant blindness, and how the heat generated can actually pop a balloon, and that because of the dark color of the inner balloon, it will actually absorb the light and heat of a laser at a faster pace. Talk about how laser pointers are also the best cat toy ever invented! Ha! Remind them that this balloon has been in their view the entire time during the performance. "You have a randomly selected color in your mind, right? Concentrate on that color in your mind." Hand the spectator the ray gun.

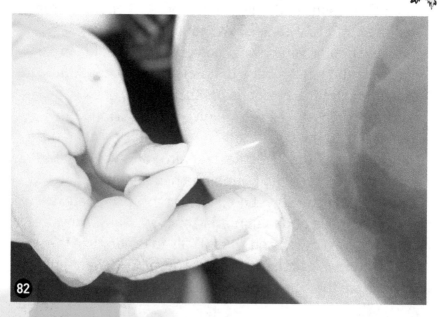

82

- Pinch the knot of the clear balloon with your right thumb and index finger. Your open left hand gently presses against the other end of the balloon. Use your middle, ring, and pinky fingers from your right hand to push against the balloon a little as your index finger and thumb hold the pin pulled back (Figure ❽❷). This position allows you to gently press and hold the black balloon without yet popping it with the needle. Let's call this the "popping position." To pop the black balloon, all you will have to do is slowly bring the needle inward with your index finger and thumb. That's the secret! Instead of depending on a dangerous laser pointer in performance, we mimic the science using a little sleight of hand. Now, back to the performance!

- Hold the balloon in "popping position," your left side toward the audience, with the spectator standing a few feet across from you. You position yourself so that you are not directly facing the spectator, but slightly turned (Figure ❽❸). This way, the right hand that has the needle

is never visible to the audience or the spectator holding the ray gun.

- Once in position, remind the spectator that this is seriously dangerous! Instruct them to keep the ray gun pointed toward the floor with their finger off the trigger until you give them the go-ahead, and then and only then should they raise the gun, pointing it toward the very center of the black balloon and nowhere else, and push the trigger button. Once the emphasis of safety has been established, and they are still pointing the ray gun toward the floor, you can continue.

- Say, "You have your random color in your mind, correct? What was your color? Say it aloud for the first time!" They say, "Red," and you immediately say, "Magic is science! Shoot and hold the laser beam at the black balloon (Figure **84**)!" The laser is triggered. Wait a few seconds! Then, slowly bring the needle inward as you press the black balloon between both hands (Figure **85**). POP! The red confetti spreads throughout the clear balloon! Show the spectator that the color matches from the one they chose! With your left side still facing the audience, slowly pull the sewing pin out with your right index finger and thumb. Once out, hold it flat in a finger palm position as you hand the spectator the balloon as a souvenir! Everything ends clean and the giveaway makes it all even more memorable.

Learn this routine! Perform it! Take elements of it and create your own version! Safety is key. The laser we use works with just 6V coin cell batteries for a reason. The whole routine was created based on safety. I didn't want to take any chances with a more powerful laser. Using the sewing pin to pop the inner balloon makes this something that can be used without the laser too... you could pop the black balloon with just your mind! Get creative with the principles at play! And remember, the thought that "maybe it is possible" is a powerful tool for the performer and helps with the misdirection of any routine we create.

#10
CHOMPER BOT

THE EFFECT:

The performer presents a monster robot named Chomper Bot, with three cards displayed in front of the robot. The cards depict three images: an ice cream cone, a cupcake, and a slice of pizza. The performer explains that the Chomper Bot absolutely *loves* one of these three items. The spectator chooses one and names it aloud. The performer turns each card over. The back of each card is blank except for the one the spectator chose... the back of that card, in big letters, reads "EAT ME!" The performer takes all three cards in a pile and feeds them to the robot. Chomper Bot comes alive, and true to his name, chomps at the cards. The performer removes them from his mouth, spreads out the cards again, and to everyone's surprise, the drawing on the selected card now has a bite taken out of it!

MATERIALS & TOOLS:

- **1 sheet of cardboard** (a pizza box works well!)
- **1 red crepe party roll**
- **1 brad** (aka paper fastener)
- **1 Arduino UNO**
- **2 micro servos**
- **1 9V battery plug**
- **1 9V battery**
- **2 sheets of blank white 8.5" × 11" paper**
- **6 male-to-male jumper wires (or 22 gauge solid core wire)**
- **1 repositionable/re-stickable glue stick** (this is a type of glue that creates a sticky but not permanent bond, allowing us to stick two items together, and then easily remove them and re-stick again. Elmer's and Scotch brands both made good versions)
- **9 unlined index cards**
- **Masking tape**
- **Duct tape**
- **Craft glue stick**
- **Hot glue gun**
- **Hot glue**
- **Black permanent marker**
- **Pencil**
- **Small screwdriver**
- **Assorted markers**
- **Tape measurer**
- **Small X-Acto knife**
- **Ruler**
- **Scissors**
- **Wire cutters & strippers**

THE BUILD:

- Cut a 6" × 16" piece of cardboard (Figure ❶).

- With your X-Acto knife, lightly score three straight lines 4" apart from each other, across the short length of the cardboard (Figures ❷ and ❸).

- Fold the cardboard at the lines to create a 3D rectangular structure, securing the inside seam with masking tape (Figure ❹).

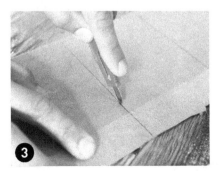

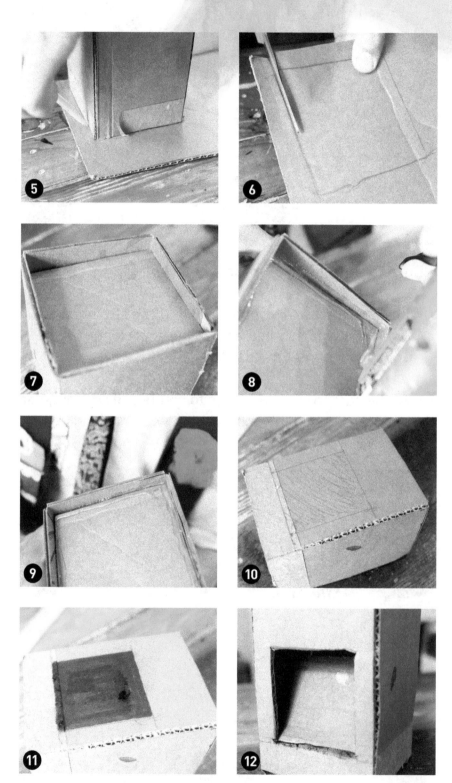

- With a pencil, trace the outside of one of the square ends of the 3D rectangle onto another piece of cardboard (Figure ❺).

- Cut the square shape out just along the inside of the traced line, then trim and adjust the square so that it can fit into the 3D rectangle and stay firmly in place, about ½" down from the opening (Figures ❻ and ❼).

- Put an even line of hot glue along the edges, then before the glue has dried, slide the square about ⅛" up toward the opening, pushing from the inside. This will help hold it securely in place. Let the glue dry completely before moving onto the next step (Figures ❽ and ❾).

- It's time to make Chomper Bot's mouth! Lay the box on its side, with the closed end facing away from you. With your pencil, draw a roughly 3" × 3" square on one side of the box, starting about an inch from the open end and ½" from either side (Figure ❿).

- Color in the square completely with a black marker (Figure ⓫).

- With your X-Acto knife, cut the top and side lines of your mouth only, leaving the bottom line — closest to the open end — completely attached. Bend the black square inward, making the fold line at the very bottom edge of the black square (Figure ⓬).

- Cut a piece of red crepe party paper about 5" long. With your scissors, round off one end, then outline the entire thing with a black marker around the edges. Also, draw a line lengthwise down the middle. This is Chomper

Bot's tongue (Figure **13**)! If you don't have crepe party paper on hand, you can make a tongue from regular paper instead! Just cut out a 2" × 5" piece of paper, color it in red crayon, and continue as usual with the black marker outline.

- Tape the non-rounded end of the red tongue to the inner side of the black square, and drape the tongue over the square, so it sticks out (Figure **14**).

- Next, we make our teeth and gums! Cut two 8.5" × 1" pieces of cardboard (Figure **15**).

- Use a craft glue stick to adhere plain white paper to one side of each piece of cardboard. Trim the excess (Figure **16**).

- Draw and/or cut out a teeth pattern on the paper side of the cardboard strips. Get creative! Maybe your Chomper Bot has

fangs (Figures **17** and **18**)!

- Put a line of glue stick above the teeth and adhere a strip of red crepe party paper to create Chomper Bot's gums. Fold it over and glue it down on the other side as well, trimming any excess (Figure **19**). If you don't have red crepe party paper you can color in gums with a red marker!

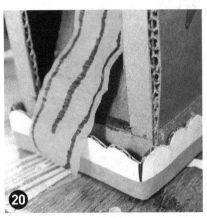

- Mount one set of teeth to the bottom part of the mouth, folding the ends so that the teeth sit ½" out from the cardboard face and are equal on each side. Put a dab of hot glue on each end to secure it in place (Figure **20**).

- With your pencil, make a mark on the left side of the Chomper Bot, when it's facing you, 3¼" up and ½" back from the front (Figure **21**).

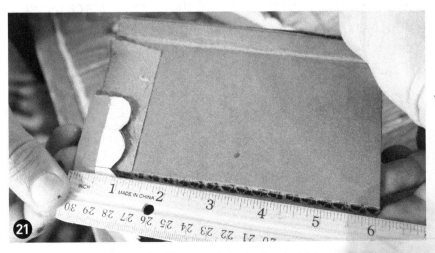

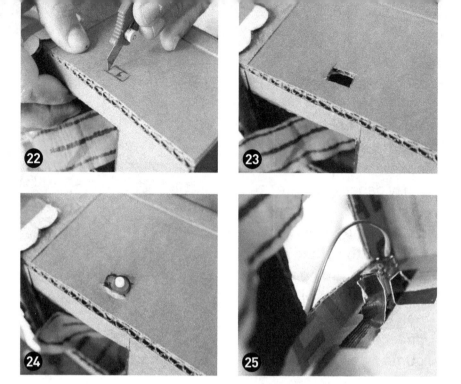

- On the mark, use your X-Acto knife to cut a small ¼" × ½" rectangle (Figures **22** and **23**).

- Push a micro servo shaft through that small opening from the inside, with the servo wires sticking straight upwards toward the top of the box (Figure **24**).

- Make sure the black square of the mouth is pushed down and out of the way, then mount the servo in place using duct tape. A small piece of duct tape pressed from the servo to the cardboard on each side will help secure it (Figure **25**).

NOTE: I often use duct tape in projects, so that I can later re-use servos for other projects, but if you'd like it to be more permanent, then you may absolutely choose to carefully hot glue the servo in place.

- On the same side of the box that the servo is mounted, make a mark with your pencil 2¼" away from the servo shaft opening, toward the back of the box (Figure **26**).

- On the mark, cut a ½" slit. Carefully thread the servo wires through the slit from the inside out (Figure **27**).

- Next, we want to make our mouth chomp! To do that, we need to connect the servomotor to Chomper Bot's upper set of teeth. Lay the teeth on a flat work table, so that the backside is facing up, and the gums and teeth are against the table. The teeth end should be pointed toward you. With your X-Acto knife, make a ⅜" hole about ½" in on the last tooth on the left side (Figure **28**).

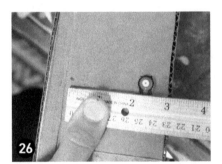
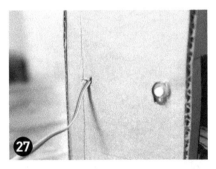

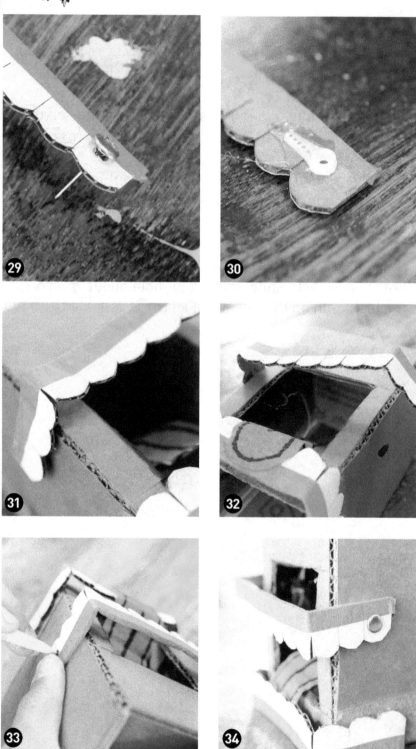

- Turn the piece around, so the gums and teeth are facing you, and insert a brad into the hole. The brad should be able to slide in easily and turn without catching. If it is tight or constricted, open up the hole a tiny bit more and try again (Figure **29**).

- Turn the teeth back around, so the gums and teeth are facing the table, with the teeth pointing toward you. On the opposite end of the teeth from where you mounted the brad, in the same position as the brad but on the right side, hot glue a one-armed servo horn, with the arm pointing straight left. Make sure to hot glue only the end of the servo arm, and do not allow any glue to block the center hole of the servo, where it attaches to the servo shaft (Figure **30**).

- After the glue dries completely, push the servo horn onto the shaft of the servo, being sure to hold the servo — not the box — as you secure it in place (Figure **31**).

- Carefully fold the upper teeth into place, making one fold approx. 2" from the end that's attached to the servo, and another fold approx. 2" from the other end (Figure **32**).

- On the other end, align the teeth so they line up straight and maintain an equal distance from the box the entire way, and mark the box with your pencil where the brad opening meets the box (Figure **33**).

- On the mark, carefully cut out a hole to fit the brad, and check to ensure that the brad can easily push in and turn freely through the opening. Close the brad loosely but securely, to allow for free movement (Figure **34**).

• Our second servo gets mounted at the top of the box to control Chomper Bot's eyeball! Position it at the top front center of the box, with the servo shaft facing forward and slightly extended over the edge. Cut or rip a 4" long and ½" wide piece of duct tape to tape the servo in position. The servo will naturally tilt back a little as you press the tape down. Use a few more small pieces of duct tape to secure the sides and back (Figures 35 and 36).

- Mount the Arduino to the top right back corner of the box with a loop of duct tape (Figure **37**).

- With your black marker, draw a 3" circle on an index card. Cut it out. With coloring tools, make it into Chomper Bot's eyeball (Figure **38**)!

- Use hot glue or masking tape to attach another one-or-two-armed servo horn to the backside of the eyeball. It should be positioned horizontally, and near one edge. When using the hot glue, again, be sure to avoid blocking the part of the servo horn that attaches to the servo shaft (Figure **39**).

- Take out your male-to-male wires or cut six pieces of 22 gauge solid core wire. You should have two red, two green, and two black 4" wires. Strip ¼" of the coating from all ends.

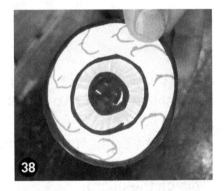

- Let's connect the teeth servo to the Arduino! Connect the brown servo wire to one end of the black male-to-male wire. Connect the other end of the black wire to any GND pin on the Arduino.

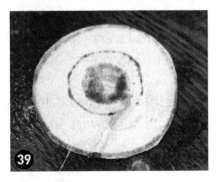

- Use the red male-to-male wire to connect the red servo wire to the 5v pin on the board.

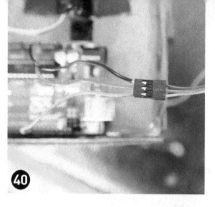

- Use the green male-to-male wire to connect the orange (signal) servo wire to digital pin 8 (Figures **40**, **41**, and **42**).

- Now let's connect the eye servo! Use a black male-to-male wire to connect the brown servo wire to any GND pin on the board.

- Use a red male-to-male wire to connect the red servo wire to 3.3v on the board.

- Use the green male-to-male wire to connect the orange (signal) servo wire to digital pin 7 on the board (Figures **43** and **44**).

- Now, before uploading the code, separate the teeth from the servo by carefully popping the servo horn off the servo (Figure **45**). We do this because we need to upload

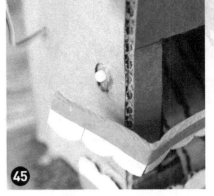

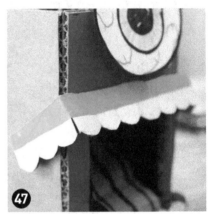

our code to get our servo in the right position first, before reattaching.

- Open up your Arduino software on your computer, and copy and upload the code found at mariothemagician. com/robotmagicchapter10 to your sketch.

- Once uploaded, you should hear both servos move slightly into position. Once you hear that, unplug the board from your computer. (If you don't hear any servo movement, go back and check that your servomotors are all connected correctly and that your code is correct.)

- Attach the eyeball to the top servo (Figure **46**).

- Attach the teeth to the side servo, so that the teeth are positioned as open as possible (Figure **47**).

- Once both are mounted, let's test it out! Plug in the Arduino and watch! After two seconds **(delay(2000))** the eyeball will start moving left to right (four times.) Then, after three seconds **(delay(3000))** the mouth will begin to chomp as the eyeball wiggles! Once we know all of this is working, disconnect the USB cable from the Arduino, plug a 9V battery into the board, and test once more.

- Then, duct tape the battery to the top, mount the plug onto the battery, and set your Chomper Bot aside as we create our magic cards (Figure **48**).

- Take out eight blank index cards. Glue two of them together with craft glue stick, creating one thicker card. Take your time to make it as square and flush as possible (Figures **49** and **50**).

48

49

50

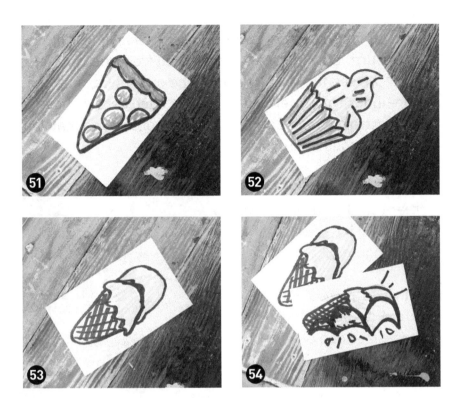

- Repeat three more times, so we end up with four double-thickness index cards in total.

- With your pencil, lightly draw a large slice of pepperoni pizza on one card. Once it looks good, outline it with your permanent black marker and color it in (Figure **51**).

- One another card, draw and color in a cupcake (Figure **52**).

- The last two cards are our magic cards. On one of them, draw and color in an ice cream cone with 2 scoops of ice cream (Figure **53**).

- On the final card, draw an exact duplicate of the ice cream cone BUT with a giant bite through it (Figure **54**)!

- Once all four cards are done, take a close look at them to make sure you can't see through at all from the backside. If you can see the image through the other side, glue one more index card layer on the back of each card to make them extra opaque. Carefully trim any excess, so that you have four solid index cards.

- Turn over the unbitten ice cream card, and write EAT ME! in big letters on the back (Figure 55).

- With your re-positionable glue stick, put a small dime-sized dab of glue on the top left corner of the front of your ice cream card with the bite in it (Figure 56).

- Put a half dollar size of re-positionable glue in the center of that same card on the back (Figure 57).

- On the card that has a full, unbitten ice cream cone, put a half dollar size of re-positionable glue in the center on the face of the card (Figure 58). Re-positionable glue is amazing, because it's... yes, you guessed it...

59

60

re-positionable! You'll soon see why this is so helpful for this trick.

- Take your bitten ice cream card, and place it face up under the cupcake card (Figure **59**). Press the cards together at the top left corner. Once they are together, carefully handle both cards as one (Figure **60**). You can gently drop it on a table and pick it up and it should stay together. To separate them, simply push the two cards apart with your fingers at the bottom right corner. This is your magic gimmick! Remember this simple move for later on. For now, press the bitten card to the cupcake again, so they stay together.

THE PERFORMANCE:

As you will understand by now, we have to make our audience believe that they have a free selection in our routine, but we will actually "force" the ice cream card on them. That said, the ice cream card is the chosen force for a reason... I have performed this routine many times, and I've found that when I ask a child to choose between these three cards, 95% of the time, they will choose the ice cream. And of course, when that happens, you don't need to force a card at all, and the impact of the trick becomes even stronger. But what if they don't pick ice cream? This is where something called the P.A.T.E.O. (Pick Any Two, Eliminate One) force in magic, invented by Roy Baker in 1968, comes in very handy! Here's how to use his method for our purposes:

The performer introduces three picture cards... a cupcake, an ice cream cone, and a slice of pizza. They also introduce a hungry robot named Chomper Bot. Chomper Bot is so very hungry (Figure **61**)! A spectator is asked to name one of the three cards aloud. If they name ice cream, you immediately prove you knew it by turning over each of the other two cards, emphasizing their blank backs. You turn over the ice cream card to reveal a big EAT ME (Figures **62** and **63**)!

Let's say they don't name ice cream, but say pizza instead. You say, "Awesome! Name another! Do this all without skipping a beat. If they select the cupcake next, pick up the cupcake card, too, so you have the pizza card in one hand and the cupcake in the other. You then ask one last time, "Are you sure? You can change your mind!" If they are sure, then at that moment, shout, "That's crazy! You could have left any card on the table!" Flip over both the cupcake and pizza cards, revealing their blank backs. Say, "The only card left, the only one out of all three!" Turn over the ice cream card, and say, "It says EAT ME!"

Now, what if they name ice cream as the second card? Well, then you say, "Perfect!" Hold one card in each hand, and without skipping a beat you say, "Pizza and ice cream! Pick one!" If they say "ice cream" you go straight to the ending from there. If they say "pizza," you quickly put the pizza down next to the cupcake, leaving you with the ice cream in your hand, and go from there. This kind of patter takes practice, practice, practice! It's amazing how good you can get at this and how convincing you can be, bringing every free decision smoothly right to the same outcome. This force is worth mastering, because it can be adapted to so many magic routines!

Now, after forcing the ice cream card and showing the words EAT ME, it's time to bring out Chomper Bot. Say, "Chomper Bot is hungry! And it's a good thing you chose ice cream, because Chomper Bot *only* eats ice cream! Watch!"

Stack the cards together, one by one, face up in your left hand, starting with the ice cream card. Place the cupcake card on top of the ice cream card (Figure **64**), and the pizza

card on top of the cupcake card (Figure **65**). This order is vital! First the face up ice cream, then the face up cupcake, and last the face up pizza. Squeeze the center of the cards together and turn the cards over as you put the stack into Chomper Bot's mouth (Figures **66** and **67**).

Plug the 9V battery into the Arduino, as you say, "In a moment, Chomper Bot will chomp these cards! But only one card is Chomper Bots favorite! The one you chose! The ice cream!"

Chomper Bot's eye starts moving, and his mouth starts chomping (Figure 68)! After the chomping stops you take the stack out. Turn the cards face down showing the "EAT ME!" words written on the bottom card. Thumbs on top and fingers underneath you gently separate the cards face down in front of you in a row (Figures 69 and 70).

Then, turn over each card slowly, making sure the EAT ME card is the last one turned. As you turn each card, say, "Look nothing happened to the pizza! Look! Nothing happened to the cupcake! Look... the one card that says

EAT ME! - the card you chose - is the only one that has changed!" Turn over the final card, and now the ice cream has a bite right through it! All three cards are cleanly shown, but the ice cream card has completely changed (Figure **71**)!

Let's take a look back behind the scenes again. The bitten ice cream card starts behind the cupcake card in the beginning. Remember, there is only a very small amount of glue between them in one corner. So, when we stack all the cards together in the right order and squeeze them together in the center as we put them in Chomper Bot's mouth, the larger amount of glue between the back of the bitten ice cream card and the front of the full ice cream card stick together even more! Thus, when we slide the cards apart at the end, we will have a visible bitten ice cream cone. This is a great example of how to visibly change cards into other cards before a spectator's eyes.

Try making other kinds of drawings or cards change! Find a theme and a storyline that fits your personality, your passion, whatever you are excited about. Change Chomper Bot to an animal or robot that suits your version of the routine. Start with a drawing and cut it out in cardboard, piece by piece. Use the servos to make your custom creation move! Change the Arduino code, so that it moves along with your unique story. This routine works really well on its own, but it's also a good base to create many different routines. My heart's goal for this book is to teach you strong magic principles and ideas of how to connect them to your maker projects, to help you see how simple robotic movements can lead to something that can make magic. This is what being a maker magician is about.

#11
PEPPER'S GHOST IN A CEREAL BOX

Pepper's Ghost is an insanely visual magic technique used in all kinds of theater, circus sideshows, haunted houses... even Disneyland uses this effect! It's also been used to bring people "back from the dead". Yes! Famous artists like Michael Jackson and Tupac Shakur, have been recreated virtually and then projected into 3D on stage to a live audience. The illusion can be *that* real looking when executed correctly. The illusion was popularized by John Henry Pepper, hence the name "Pepper's Ghost". Here, we will explore the principles of this illusion using the Arduino and the power of PWM (Pulse Width Modulation.) But we will be making this illusion in miniature, all built within a cereal box and with LEGOs!

THE EFFECT:

Inside of a cereal box is a tiny lit-up LEGO scene. A small bouquet of flowers is seen next to a LEGO mini figure. There's a small button on top of the box. When the button is pressed, the flowers slowly dissolve away until they have completely vanished! When the button is pressed again, the flowers slowly re-materialize right before your eyes. The illusion can be repeated over and over again.

How does it work? The whole scene is visible through a small clear LEGO window inside the box. LEDs are mounted inside the box, lighting up two separate spaces. One is the tiny room that we see throughout, and the other is a space completely hidden from the audience. The hidden space has flowers inside of a vase. The visible space has just a vase with no flowers. Now, when that hidden space inside the box is illuminated, it creates a strong visible reflection onto the window. To the viewer, that reflection makes the flowers seemingly appear and disappear inside the empty vase. Because of the lighting and the viewer's angle, to the naked eye, the flowers fit in perfectly with the other items in the room. There is no distinguishable difference, until the LED lights up from the hidden space, and the flowers slowly fade away. As they fade, the LED from the visible room slowly lights up. This switch of LEDs from room to room is what creates this illusion. This is the part I love most... using the PWM pins on the Arduino, the LED fades in such a way that the flowers almost look computer generated! They slowly deteriorate into nothing, and then re-appear like magic.

This project teaches you how to use LEDs and programming to create magic. Plus, you have an excuse to get your favorite cereal! I promise, if you build this once, you will recreate it over and over again. This is the kind of illusion that can be adapted in so many different ways, and you can experiment with so many different objects!

Let's get onto the build!

MATERIALS & TOOLS:

- 1 family size cereal box
- 1 pizza box or scrap cardboard
- Assorted structural LEGOs
- Small LEGO flowers
- 1 LEGO mini figure
- 1 clear plastic LEGO panel, 1.75" × 1.75" (trans clear panel wall)
- 2 white LEDs
- 2 270 Ω resistors
- 1 10K resistor
- 1 momentary button
- 1 Arduino UNO
- 1 9V battery plug
- 1 9V battery
- 1 spool of 0.6 mm solder wire
- 22 gauge solid core wire (black, green, red)
- Hot glue gun
- Hot glue
- Glue stick
- Soldering iron
- Wire strippers
- Wire cutters
- Scissors
- Ruler
- Tape measurer
- Masking tape
- Duct tape
- Pencil
- Small X-Acto knife
- Black permanent marker

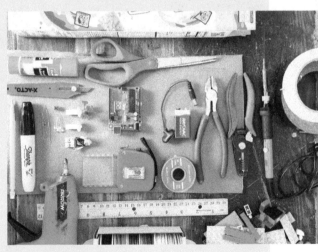

THE BUILD:

- Place your cereal box flat on its back, with its front cover facing you. Measure, mark, and cut a straight horizontal line 8" from the bottom of the box on the front, left, and back sides (Figures ❶ and ❷).

- On the right side of the box, measure, mark, and cut a horizontal line 12" from the bottom of the box. Cut the edges down to the 8" lines cut before. You should now have an open-topped cereal box shortened to 8" all around except for on the right side, which should extend 4" above the rest. Set aside the scrap pieces for later (Figures ❸ and ❹).

- Place the box flat on its back, with the opening facing you.

- Cut right along the corner edges on both sides of the face of the box, starting from the opening, and cutting all the way down to the bottom of the box (Figure ❺).

- Bend back the large front flap we've created, so that it stays open (Figure ❻). This inside area is where our rooms will be built.

- We need to divide the floor of this area into four squares. Do this by measuring and making two pencil lines, dividing the floor in half horizontally and vertically (Figure ❼).

- With both openings still facing you, draw a very faint diagonal line in the floor's front right square, the square to the right and closest to you. The diagonal line should start at the center of the floor, where the four squares meet, and end on the opposite corner, on the edge of

the opening (Figure **8**). This angled line is where our reflection will later take place.

- Now, the back left square on the floor needs to be hidden from view for performance. Cut two pieces from the discarded cereal box cardboard to create walls for that space. Dimensions will vary depending on the size of your cereal box.

- Use those pieces to wall in that back left square, securing the pieces in place with masking tape or lines of hot glue. Make sure the graphics of the cereal box are facing the inside of the square (Figure **9**).

- Now, fold the extended 4" piece of cereal box inward, so that it's creating a partial wall on the front outer opening. If it extends past the centerline, trim it down. Use a thin line of hot glue or masking tape along the bottom edge to hold it in place (Figures **10** and **11**).

- With LEGOs, put together two roughly 3.75" long × 3" wide × 2.5" tall (dimensions will vary depending on the

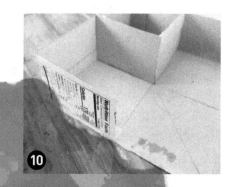

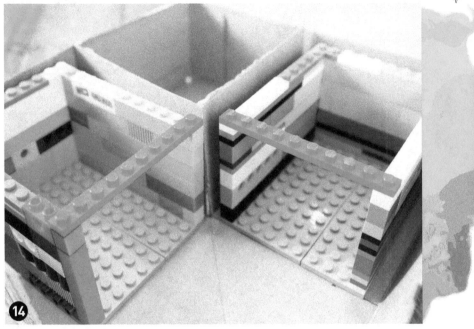

size of your cereal box) rectangular rooms, each with a floor, two long side walls, and one short side wall, but no ceiling. It's important that the flooring match in both rooms. I used light grey LEGO panels for my flooring. The walls can be different from room to room (Figure 12).

- To ensure a good fit, insert the LEGO rooms into the back right and front left squares of the cereal box structure, with the unwalled sides of the rooms facing outward (Figure 13). Make any adjustments to fit as needed.

- On the sides with no walls, add a long, thin LEGO piece across the top, creating a top to a doorway of sorts (Figure 14).

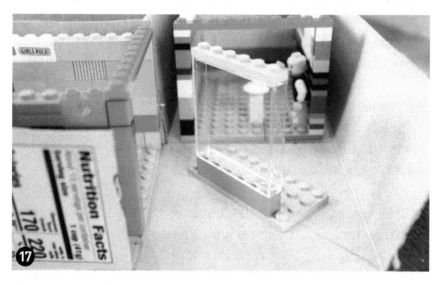

- We will also need a LEGO mini figure, the flowers from a small LEGO flower pot, and two small matching white LEGO pedestals. Take your small pedestal with flowers and place it in the front left LEGO room. Place the other pedestal, without flowers, front and center in the back right LEGO room. Place the LEGO mini figure in the back right room, too, with its back against the right wall, right in the front. We'll adjust exact placement for everything later (Figures ⑮ and ⑯).

- We need a clear plastic lego window (trans clear panel wall). The one I used is about 1¾" × 1¾" in dimension. Press this window panel onto a 2½" × 1¼" flat grey floor piece and additional LEGO piece or two to raise it about ½" from the floor. Add a LEGO piece or two to the top as well, so the height is the same as the height of the cereal box walls (Figure ⑰).

- Eventually, we will place the window panel structure along the diagonal line in the front right room. This will be the viewing window to the illusion. But for now, set it aside.

- Let's get to work on the electronics! Plug in your soldering iron. Take out two LEDs, and cut four 8" 22 gauge wires — two green and two black — with ¼" of the coating stripped from all ends.

- Taking a look at our LEDs, we see that each LED has two wires. The longer wire is the positive wire, and the shorter wire is ground. This is important to remember! If we plug in an LED the wrong way, it won't work!

- Tape down the LEDs onto a safe surface to solder on, and

solder a 270 Ω resistor to the longer (positive) wire of each LED.

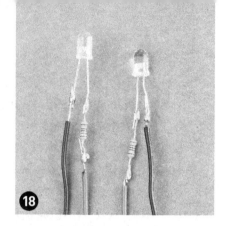

- Solder a black wire to the shorter (ground) wire of each LED (Figure **18**).

Set your soldered LEDs aside, and take out your momentary button and 10K resistor.

- Cut three 22 gauge wires — one green 8" wire, one black 6" wire, and one red 6" wire. Strip ¼" of the insulation off all of the ends.

- Carefully spread out the wires from the momentary button, so that the bottom of the button can lay flat. Tape your button down on one side with a piece of masking tape. Notice that the button has two wires on each side. Put a small drop of solder on each wire on just one side of the button. Don't spend too much time doing this, though... we don't want to weaken the plastic housing from the button by pressing in for too long. A few seconds heating each wire with solder should do.

- Take one end of your 10K resistor and wrap it around one of the soldered pieces. We want the resistor resting right up next to the button's wire. Solder it on!

- Wrap the wire from the other end of the resistor to the black wire.

- Bend and adjust so everything points straight out ,and solder it in place. The wire from the resistor is very fragile! If you have it on hand, add shrink tubing reinforcement or a tight, small, rolled piece of masking tape. This will help strengthen the wire so we can bend it later on with less risk of breakage.

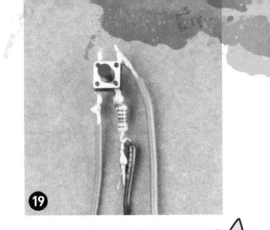

- On the other side of the momentary button, solder the wire directly across from the resistor to the 8" green wire. Keeping all the wires pointing in the same direction, lay your green wire along the black one first, then solder it to the button's wire.

- Solder the 6" red wire to the other wire on the button that we pre-soldered, right next to the wire with the resistor (Figure **19**).

- We should now have all three wires pointing out in the same direction. We want the button facing up and the wires all pointing out to one side. Again, remember that the connections to the button are very fragile! Over-bending the wires from the button could cause the wires from the button to snap. Let's reinforce it:

  - Plug in your hot glue gun. Put a small piece of masking tape, sticky side facing up, underneath the connection points, to catch the hot glue (Figure **20**).

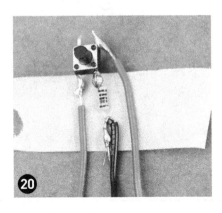

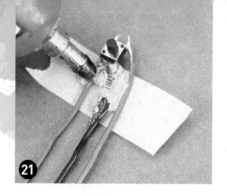

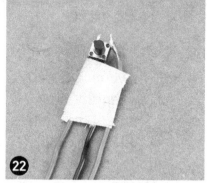

- Apply a very small amount of hot glue to the connection points (Figure **21**). Don't overdo it!

- Once the glue has dried, fold the tape over the hot glued connection points, sealing and protecting the weakest parts of the wires (Figure **22**). Now, we can move on to mounting and connecting to our board!

- With a loop of duct tape, mount the Arduino board to the floor of the back left room in the cereal box, with the USB and battery ports facing the back (Figure **23**).

- We need to mount one of the LEDs to the top "doorway threshold" of the front left LEGO room — the room with the flowers in it. The LED should be positioned so that it's pointing downward into the center of the room, right toward the flowers. A thin strip of duct tape across the thin LEGO bar on top will hold the LED in place (Figure **24**).

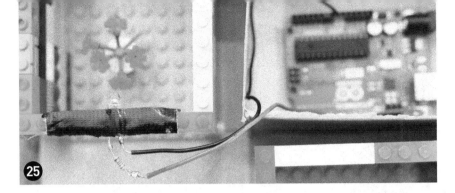

- Carefully bend the black wire from the LED back and down towards the Arduino, plugging it into any GND pin on the board. Carefully bend the green wire back and down toward the Arduino, too. Make sure the metal from the green wire never touches the black wire! Plug the green wire to digital pin 11 on the board (Figures **25** and **26**).

- With a thin strip of duct tape, mount the other LED to the top doorway threshold bar of the other LEGO room — the room with the LEGO mini figure and empty pedestal. We want the LED to be pointing downward into the room and towards the front center (Figure **27**).

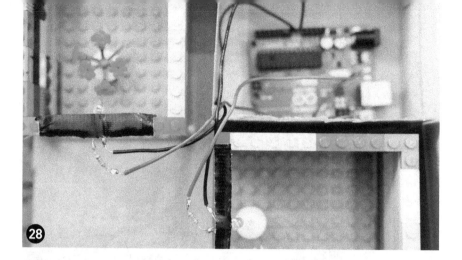

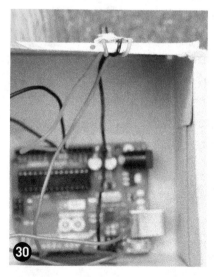

- Carefully bend the green wire across and down toward the Arduino board. Plug that green wire to digital pin 10 on the board. Carefully bend the black wire across and down toward the Arduino. Make sure the metal from the black wire never touches the metal from the green wire! Plug the black wire to the GND pin that's next to digital pin 13 on the board (Figures **28** and **29**).

- Take your momentary button out, and plug the black wire from the button to a GND pin, the red wire to the 5v pin, and the green wire to digital pin 5 on your board. Bend the wires together and fold them over the outside back left edge of the cereal box, so that the button rests pressed against the outside of the cereal box (Figures **30** and **31**).

- With your X-Acto knife, carefully cut a small door/flap on the lower back wall of the cereal box, to access the Arduino's USB and 9V plug. An opening of roughly 2" wide and 1" high is sufficient, leaving a flap so that it can be opened and closed (Figure **32**).

32

- Open up the Arduino software on your computer, and plug your board in! Copy and upload the sketch from mariothemagician.com/robotmagicchapter11.

If the sketch doesn't compile, make sure each and every line of code, including the " **}** " symbols, are all copied exactly. Now, let's test!

Once the sketch is uploaded, there will be one LED that is on. The room with the flowers in it should be lit up. Press the button. The LED will slowly start fading out as the other LED starts slowly fading on. If all works as described, we can move on to fine tuning! (If it doesn't, check all wire connections, LED connections, button connections, and connections to the board, then try again.)

Once we know the button is triggering the proper LED functions, we need to fine tune our set up to create the illusion:

- Place your LEGO window panel so that it is aligned and centered along the diagonal pencil line drawn in the front right room of the cereal box (Figure **33**).

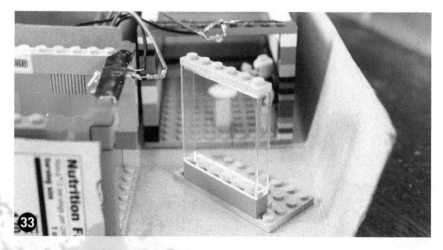

- Plug your 9V battery into the Arduino board through the back flap on the cereal box, and mount a 9V battery right outside the back of the box, where the plug can easily reach it. This is so we can turn the Arduino on and off by popping the battery off of the plug rather than pulling the plug directly from the board. The less we disturb the mounted wires from the LED and button, the better (Figure 34).

- Let's hone the position of the pedestals and flowers. The bumps on a LEGO piece are called studs. The pedestal with the flowers should be positioned on the 6th row of studs from the front, in the center, in the front left LEGO room. Make sure the flowers are positioned so that at least one of them is facing directly toward the doorway (Figure 35). The LED is already bent down so that the light shines on the flowers.

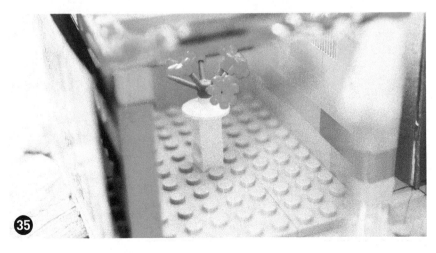

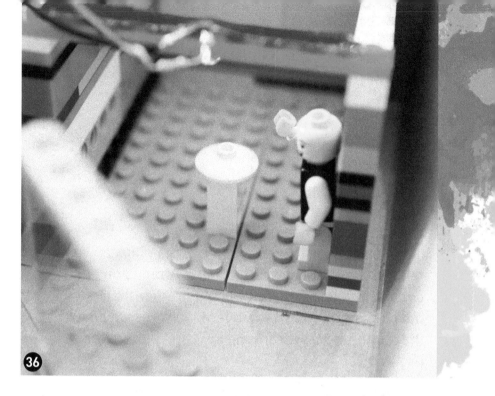

- The other LEGO room should have the empty pedestal mounted on the 3rd row from the front, in the center. The LED shines light directly toward the pedestal. The LEGO mini figure should be positioned against the right side wall, up front, facing toward the pedestal (Figure ③⑥).

- Plug in the 9V battery to the 9V plug, and close the top of the cereal box.

- Look through the front opening of the cereal box, and adjust the LEGO window panel so that you can see the reflection of the flowers and pedestal through the window (Figure ③⑦).

- Continue to adjust the window until you find the sweet spot where both pedestals are synced perfectly, without any image doubling, and the flowers appear to be right there in the room with the LEGO mini figure (Figure ③⑧).

- Adjust the LED from the room with the actual LEGO flowers, so that the flowers appear bright and clear through the window.

- Once you have it all synced, press the momentary button.

- Watch through the window as the flowers very slowly vanish (Figures **39**, **40**, **41**, **42**, **43**, and **44**)!

- Press the button again, and watch them slowly re-materialize! Such a beautiful illusion!

To make your build sturdier, you may choose to tape down the LEGO rooms in the cereal box. You may do the same with the window, once you've found the right position. You may also choose to add another cardboard panel to the front opening of the box so that the only thing visible through the opening is the LEGO window (Figures **45** and **46**). That can make this illusion into something that really fools the audience. Or, you can leave it open as is, and use this as a science demonstration of magic history and why it works! You could even make a giant life-size version to fit an actual person, made of plexiglass, light bulbs, and wood! Wherever you go with this, I hope it has inspired you to explore ideas!

Additional tips: If you'd like the illusion to happen faster, you can change the fade time in the code! Look for **delay(55);** in the sketch. One will be inside **case 1:** (vanish) and the other **delay(55);** will be inside **case: 2** (reappearance.) Changing both delays to a smaller number, like **delay(5);** will make the fading go much faster. Please experiment! Try different objects, too! Replace the flowers with a different object. Create a storyline!

You can also use this illusion to play with the concept of adding color to an object instead of making an appearance. Or, you could have two small Rubik's Cubes, one mixed, and one solved to see if you could make the Rubik's Cube seem to magically solve itself. Play around! The only way to expand this illusion is to experiment. Maybe add a moving element! Use the Arduino's digital pins to add a servo and make something else in the room move as the object vanishes and reappears.

As with everything in this book, this old illusion is made new again because of the Arduino and its capabilities, accessibility, and ease of use. I hope you will grasp the concepts in this book and take them to new creative heights!

Magicians, dig through those old magic books on your shelves, but this time look at each classic with new eyes. How could you adapt technology to reinvent them? Tinkerers, dig through your knowledge and skill set, and think about how you can use it to entertain audiences, to make someone laugh, to create a narrative or story.

My goal has been to bring two worlds together, to bring inspiration from both sides toward each other. Mixing ideas, sparking new takes on common concepts, breathing new life. I hope these simple ideas spark a whole new way of thinking for you! I hope you look at all of the other skills and passions in your life, too, and think about what new twists, combinations, and marriages you can craft with them. Don't follow just one linear path, but criss cross and zig zag and explore along the way.

I CAN'T WAIT TO SEE WHAT YOU CREATE.

ABOUT THE AUTHOR:

Mario "the Maker Magician" Marchese is a touring performer for all ages. He is recognized in the magic industry as an innovator, integrating DIY robotics with homemade magic to create original works. His act is magic through the lens of the Maker Movement, with a dedication to accessibility and hands-on learning. Mario has appeared on *Sesame Street,* NBC's Universal Kids, and live on tour with David Blaine, who calls him "the best kids' magician in the world!!"

Mario is the creator of the cardboard magic viral video sensation known as Automabot. He is also the author of the beginner DIY magic book, *The Maker Magician's Handbook*, published by Make Community LLC. Mario's works and tours are a family business through and through, and his wife, Katie, and children Gigi and Bear, are his constant collaborators and co-adventurers! For more about Mario, visit mariothemagician.com.

CPSIA information can be obtained
at www.ICGtesting.com
Printed in the USA
JSHW050319180921
18813JS00002B/3